I0503517

Entangled Magazine

Technogenetics:

Synthetic Genetic Biological Weapons (sGBW)

Volume 86 July 2024

**Published
by
Anthony Patch**

**Edited
by
Kathleen Patch**

Cover: *Haman Begging the Mercy of Esther*, by Rembrandt

Table of Contents

Table of Contents

Cast thy bread upon the waters…

Ecclesiastes 11:1

"Cast thy bread upon the waters for thou shalt find it after many days." This verse is traditionally attributed to King Solomon. Ecclesiastes is known for its philosophical reflections concerning life, attaining wisdom, and the futility of earthly pursuits.

What does the *"cast thy bread upon the waters"* verse really mean? In ancient days, *"casting bread upon the waters"* was a figure of speech relating to displays of kindness and acts of charity. To cast your "bread" meant to spread generously abroad - taking from your own resources and abundance, selflessly, without expecting anything in return. This led to the understanding that in time, God – who sees all, would reward such kindness.

But what if the "cast thy bread…" phrase had another meaning – one quite the opposite from what God intended? What if it was satanic in nature and nefarious in design? Further, what can we expect to happen to those who take from God's Holy Word and apply it to their own catastrophic, biogenetic weaponry, turning this Biblical altruism into a murderous act bringing about widespread death and desolation?

Applying the Word of God to heinous projects and criminal acts is blasphemous! The anticipated "return" will not be rewarded with anything good. For certain, death of spirit, soul, and body await those who connive to bring about such annihilation. This is the ancient story of Haman, the Agagite, and, in more recent history, Adolf Hitler – combined. Both men, and their documented plans, parallel one another and proclaim the intent to abolish entire ethnic groups of people.

Those who were once in the target are now targeting others. Apparently, lessons were not learned from the past. The hunted become the hunters…it begs the question, "Who were the practitioners and teachers of these concepts?" and, more importantly, "Who taught whom…?"

"Charity," in the minds of those demonically inspired and possessed, may translate now to something like this:

"*Casting your bread*" symbolizes an act of spreading abroad a "sophisticated" biogenetic agent that will yield an "immediate" return. Trusting in our own knowledge and scientific research, we developed this deadly weapon in order to commit barbaric acts that will be rewarded by the elimination of our enemies. We trust that our unspeakable (and seemingly undetectable) acts will bring about the demise of all those we seek to target...

"Israel is working on the development of a biological "ethnic bomb" that kills on a racial basis, *The Sunday Times*, the Sunday edition of the prestigious British newspaper, wrote on November 15, 1998.

A biological weapon can be "tuned" to kill Arabs, such as Palestinian Arabs or Iraqi Arabs, but leave Jews unharmed."

– The Sunday Times – British Newspaper

"The intention is to use the ability of viruses and certain bacteria to alter the DNA inside their host's living cells. The scientists are trying to engineer deadly micro-organisms that attack only those bearing the distinctive genes."

– The Cradle.co, Oct. 2022

Text of Prime Minister Benjamin Netanyahu's speech before a joint session of Congress on July 25, 2024, as issued by his office:

My friends,

For decades, America has provided Israel with generous military assistance, and a grateful Israel has provided America with critical intelligence that saved many lives. We've jointly developed some of the most sophisticated weapons on Earth. I choose my words carefully: we've jointly developed some of the most <u>sophisticated </u>*weapons on Earth, that help protect both our countries. And we also help keep American boots off the ground while protecting our shared interests in the Middle East.*

Definition: Sophisticated

(of a machine, system, or technique)
developed to a high degree of complexity.

Oxford Languages

Israel's Secret, Illegal Biological War Against Arabs

By Kit Klarenberg , The Cradle.co Oct. 25, 2022

For decades the use of banned biological weapons during the Nakba was kept hidden in Israel's archives. Recent discoveries have shed light not only on this Zionist war crime, but also the sinister motive behind it.

In September, a highly revealing academic paper was published exposing the details of a previously hidden operation by Zionist militias during the 1948 Nakba (or "Catastrophe"), in which chemical and biological weapons were used to poison Palestinians, intervening Arab armies, and the citizens of neighboring states with typhoid, dysentery, malaria, and other diseases.

Working by stealth, Zionist militants poured vast quantities of infectious bacteria into wells and aqueducts providing villages, towns, and cities with water, in direct violation of the 1925 Geneva Protocol, which strictly prohibits "the use of bacteriological methods of warfare."

The local epidemics created by this man-made disaster greatly assisted the forcible conquest of Palestinian territory by armed Jewish militias with their capture made permanent, while hindering the progress of advancing Arab armies.

The 1948 War has been well-studied, and its impact, chiefly the permanent displacement of hundreds of thousands of Palestinians in the Nakba still reverberates today. Yet, understanding of the conflict has hitherto been incomplete.

Aside from opaque references to the biological warfare campaign in the diaries and autobiographies of Zionist leaders and militants from that era, and a 2003 academic article, the use of these illegal substances has never previously been revealed.

In an ironic twist of fate, the Zionist biological blitzkrieg was suppressed so successfully that numerous highly incriminating documents referring to the operation's name - "Cast Thy Bread," a Biblical quote from Ecclesiastes 11:1, in which Jews are directed to "cast thy bread upon the waters, for after many days you will find it again" - slipped past government censors unexpurgated (Ed. note: Having had objectionable or unsuitable matter removed.).

Evidently, even they were unaware of this war crime which followed the chemical extermination of millions of Jews, which says a lot.

It turned out that this gap in the historical record was both created and maintained intentionally. As the paper notes, a reference was made in the diaries of Israel's first prime minister David Ben-Gurion two days before war broke out on 15 May, 1948 to a Zionist militant recently spending several thousand dollars on "biological materials." However, this was censored by the Defense Ministry Press when the volumes were published in 1982.

That cover-up continues to the present day, even in the paper itself. The authors – Benny Morris of Ben-Gurion University and Benjamin Z. Kedar of The Hebrew University of Jerusalem – seem at pains to diminish the significance of "Cast Thy Bread," pointing to the relatively few casualties produced by the effort as a sign of its "ineffectiveness."

Such analysis discounts an obvious alternative interpretation, namely, that the relatively low death toll was in fact intended. This was due to the long-held Zionist objective of seizing land reserved for Arabs under the UN's 1947 partition plan – under which Mandatory Palestine would be split in half between separate Arab and Jewish states – and portions of neighboring Arab countries, without mass slaughter, and thus plausibly denied.

Reinforcing this theory, the paper reveals that the water supplies of several Arab villages, towns, and cities were targeted by Zionist militants even before the war, and that biological warfare was seen by Zionist militants at the time to have been pivotal in the permanent capture of Palestinian land and expulsion of local residents.

Take for instance the Zionist poisoning of a vital aqueduct in Kabri, a primary source of water for nearby Palestinian settlements, which the paper's authors call "the most serious and potent use" of biological weapons during the 1948 War, despite it taking place before the conflict formally began.

The historic northern city of Acre, which the UN designated part of a future Arab state, depended heavily on the aqueduct for water. The morale of its inhabitants is said by Morris and Kedar to have been "already shaky" when local supplies were poisoned, due to recent Zionist conquest of nearby Haifa, the region's capital.

That fall of the city led to much of its population fleeing and taking up residence in Acre, which was cut off from other key regional centers and neighboring Lebanon. This, combined with the impending withdrawal of the British – who were supposed to be defending Arabs from Zionist attack – led to "plummeting" spirits among civilians. The outbreak of a typhus epidemic reduced them to "a state of extreme distress," the city's mayor was quoted as saying on 3 May that year.

Fast forward to 13 days later, when Zionist forces attacked the city, issuing a brutal ultimatum unless Acre's inhabitants capitulated without resistance: "we will destroy you to the last man and utterly." Hours later, local leaders surrendered, leading to three quarters of Acre's Arab population – 13,510 civilians – being displaced forever.

The following month, a Zionist militant intelligence report concluded that artificially unleashing the epidemic in advance had contributed significantly to Acre's hasty collapse. The same review found that outbreaks of typhus and "panic induced by rumors of the spread of the disease" alike were similarly "an exacerbating factor in the evacuation" of several Palestinian areas.

On top of ensuring a low death rate then, biological weapons also made the mass purge of Palestinians appear self-initiated.

On 26 September, Zionist operatives began a wide-ranging campaign of "harassment by all means" against soldiers and civilians across Palestine and on the soil of Arab countries involved in the 1948 War. Expelling the occupants of territory earmarked for Jews by the UN, seizing the West Bank, and ensuring displaced refugees didn't return home, were all objectives of the Zionist project.

Zionist militants had for some time been targeting Arab soldiers directly with biological weapons. In late May that year, Egypt's foreign minister sent a cable to the UN secretary general announcing the recent arrest of two "Zionist agents who admitted that they had been instructed to contaminate the springs from which the Egyptian troops at Gaza draw their water supply."

The pair acknowledged having dropped typhoid and dysentery germs into nearby wells, and were found to be in possession of "several bottles containing a liquid which was discovered to contain the germs of dysentery and typhoid," as well as a "canteen containing a liquid with a high concentration of typhoid and dysentery germs."

Such high-level exposure did nothing to deter the execution of "Cast Thy Bread." In fact, further undermining the whitewashed narrative of Morris and Kedar, the targeting of neighboring Arab states continued until the final stages of the war, when Zionist victory was all but inevitable.

In the case of Lebanon, even before the campaign of "harassment by all means" commenced, Zionist agents in Beirut were scouting possible targets for sabotage operations in Lebanon, including "bridges, railway tracks, water and electricity sources." They were eager to cast the net further afield of "Cast Thy Bread."

As late as January 1949, two months before the country signed an armistice with the Zionists, militants were tasked with investigating "water sources [and] central reservoirs" in Beirut, and "supplying maps of water pipelines" in major Lebanese and Syrian towns.

After the 1948 War ended, the informal Zionist biological warfare unit became the Institute for Biological Research in Ness Ziona, central Israel. Its first director was Alexander Keynan, a former militant who was intimately involved in the planning and execution of "Cast Thy Bread." Clearly, his sterling work made him leading candidate for research into future offensive biological warfare strategies.

Quite where Keynan's investigations led, and the scale of modern Israel's biological and chemical arsenal today, isn't certain - although the country is one of just 13 out of 184 UN-recognized territories that is *not* a signatory to the 1975 Biological Weapons Convention, and one of just four states *not* to be party to the 1997 Chemical Weapons Convention.

Ominously, this may suggest that Israel's research in the field remains ongoing. It may also serve as another rationale for keeping such a tight lid on "Cast Thy Bread" as the notorious operation still has relevance to the present, which Israeli authorities wish to keep secret.

In November 1998, Britain's *Sunday Times* citing Israeli military and western intelligence sources reported that Tel Aviv was "working on a biological weapon that would harm Arabs but not Jews," by "targeting victims by ethnic origin."

"In developing their 'ethno-bomb', Israeli scientists are trying to exploit medical advances by identifying distinctive genes carried by some Arabs, then create a genetically modified bacterium or virus," the newspaper alleged.

"The intention is to use the ability of viruses and certain bacteria to alter the DNA inside their host's living cells. The scientists are trying to engineer deadly micro-organisms that attack only those bearing the distinctive genes."

The program was said to be based is based at a "biological institute" in Ness Iona, home to the Institute for Biological Research. A scientist at the site was quoted as saying his peers had "succeeded in pinpointing a particular characteristic in the genetic profile of certain Arab communities, particularly the Iraqi people," and that "the disease could be spread by spraying the organisms into the air or putting them in water supplies."

Critics denounced the Times report at the time as a "blood libel," referencing the fabricated, anti-Semitic myth that Jews murder young Christian boys in order to use their blood in religious rituals.

It is fitting then, that when on 27 May, 1948, Syria's representative to the UN read out the Egyptian cable sent to the body's secretary general on the capture of "Zionist agents" attempting to poison Egyptian troops in Gaza, his Jewish Agency counterpart charged that Cairo and Damascus had "chosen to associate themselves with the most depraved tradition of medieval anti-Semitic incitement - the charge that Jews had poisoned Christian wells."

According to The Palestine Chronicle, the recent unearthed documents are one of many historic war crimes committed against the Palestinian people by the then-emerging occupation state, yet much of the Nakba's history remains classified and is slowly resurfacing.

- Kit Klarenberg, October 25, 2022

Commentary

"Those skilled in war subdue the enemy's army without fighting hard. They capture the enemy's cities without a storming attack and overthrow his state without excessive and perpetual damage. Their aim must be to take all under heaven intact through strategic superiority."

- The Art of War by Sun Tzu

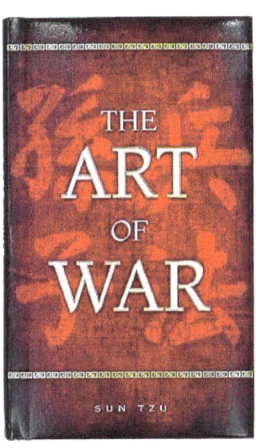

Recently, within scientific literature, emphasis has been given to concerns over bioterrorism and biowarfare involving genetically modified viruses and bacteria. Distinct from the more common forms of natural biology, *Technogenetics (*as we've coined the term and fashioned the acronym "sGBW") comprises synthetic Genetic Biological Weapons (sGBW). These possess characteristics genetically engineered to precisely target a population race, or individuals through socially transmitted non-kinetic means. These are technologically-produced, genetic bioweapons designed to induce precision microscopic injuries and/or death.

Today, the global population is confronting not only naturally occurring diseases, but additionally, the specter of bioterrorism and bioweapons comprised of genetically modified components.

These components make up novel organisms with increased virulence, infectivity, and drug resistance. Advancements in biotechnology continue to modify lifeforms. The sinister side of such advancements is referred to as "black biology" whereby novel organisms are developed and applied as genetic biological weapons.

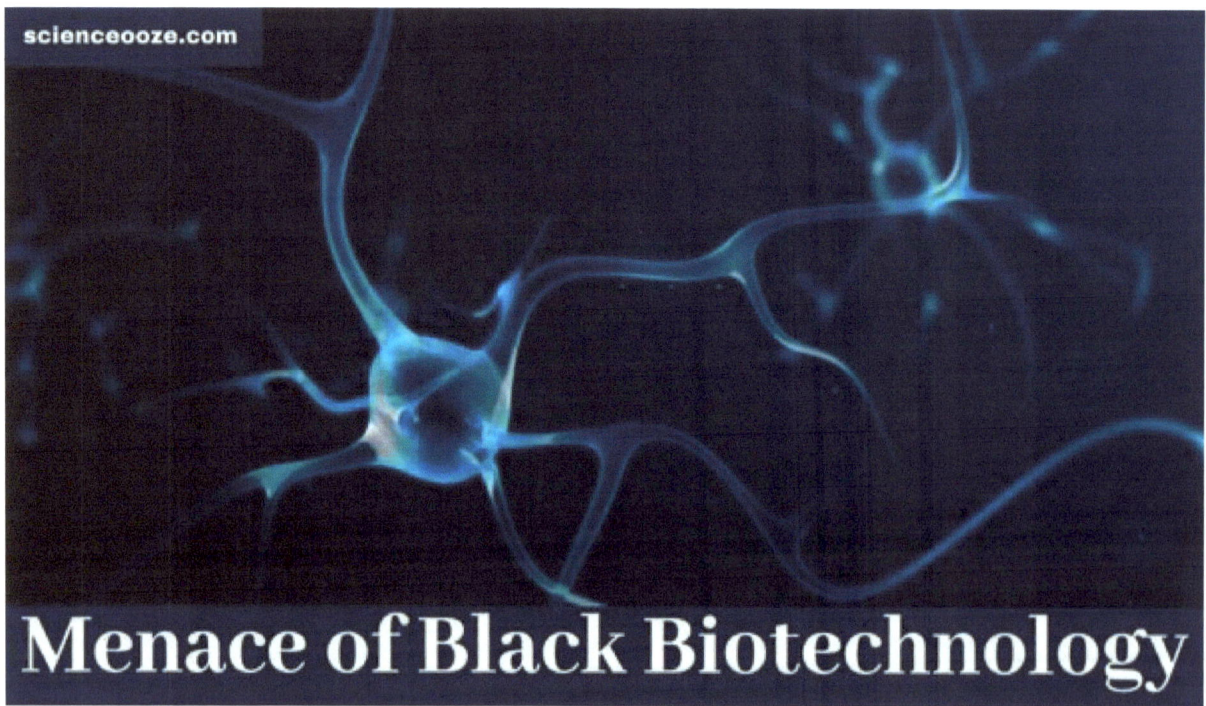

Based upon genomics and proteomics (the study of proteins and their effects), synthetic Genetic Biological Weapons (sGBW) differ from previous biological weapons (intended to damage whole tissues and organs) by targeting the specific structures of an individual gene or protein within a targeted population (race) or targeted individuals based upon the make up of their genome (the complete set of genes or genetic material present in a cell or organism).

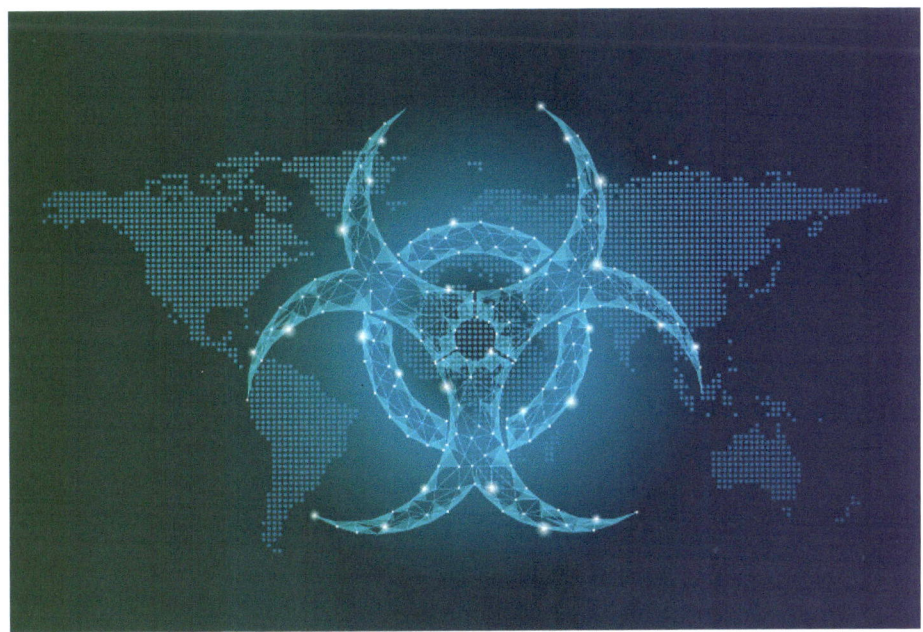

Synthetic Genetic Biological Weapons (sGBW) are designed for high levels of asymptomatic transmission targeting a vulnerable person or persons comprised of specific genes or gene sequences, as opposed to a larger population.

Populations devoid of a specific gene or sequence of genes, or by programmed obsolescence (this is when cells of a gene-induced disease undergo cell death at the conclusion of a given number of cellular generations) can be protected. The initiation and spread of the disease can be for a prescribed incubation period or designed for asymptomatic, high transmission.

The effects of sGBW can be debilitating, including death, but also mental effects impacting comprehension, learning, and even changes in personality. Additionally, these will wound in accordance with specific genes, proteins, cells, tissues, and organs. Individual targets are selected based upon a nucleotide sequence or a specific protein structure, resulting in physiological dysfunction of organs and damaged tissues.

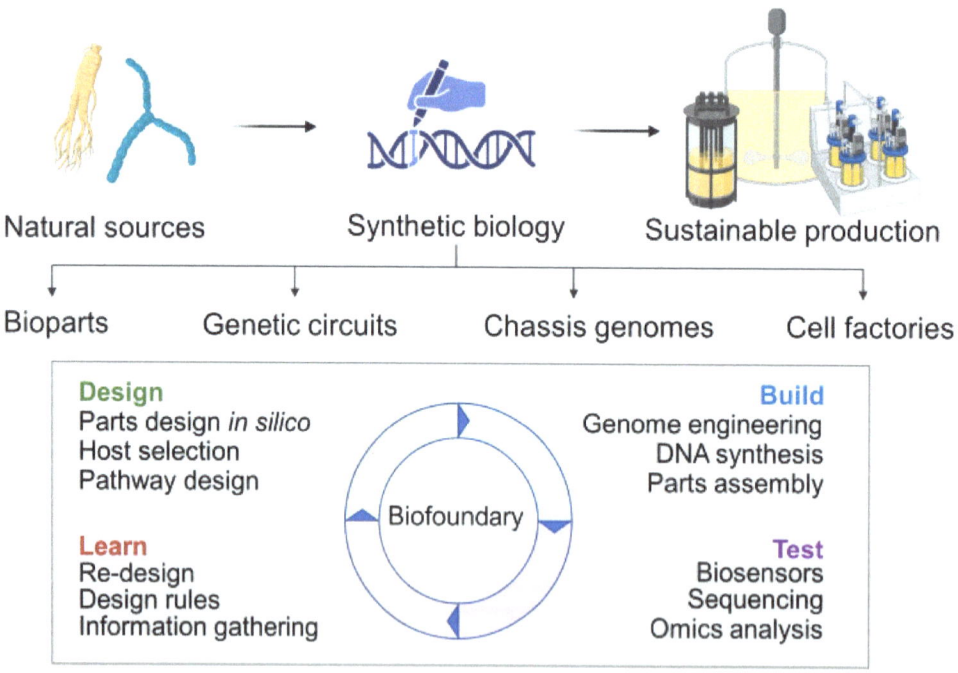

These advancements in sGBW coincide with those in *Technogenetics* overall, surpassing previous generations of genetically modified agents. Today, from a technology standpoint, it is less complicated *in silico* (in software) to design and construct a pathogen, virus, or bacteria, than to isolate one from the wild.

Synthetic biology is a dual-use technology, meaning it has the potential for both benevolent as well as malevolent uses. Specifically, the creation of unnatural components are capable of producing injuries novel from those caused by previous types of weapons, including chemical and nuclear. The latter having been designed for killing and destruction on a broad scale.

Coinciding with this, a new class of injuries, known as "precision injuries" has been realized. Designed not to kill, genetic biological damage is tailored to the specific military actions of an opposing force.

Synthetic Genetic Biological Weapons (sGBW) are having a dramatic influence on the design and manner of warfare. In the paradigm of weaponized genetic modification, only those most Technogenetically-advanced will succeed in warfare.

Genetically Engineered Bioweapons: A New Breed of Weapons for Modern Warfare

Dartmouth Undergraduate Journal of Science

By DUJS

March 10, 2013

Applied Sciences

Winter 2013

Genome sequencing has given rise to a new generation of genetically engineered bioweapons carrying the potential to change the nature of modern warfare and defense.

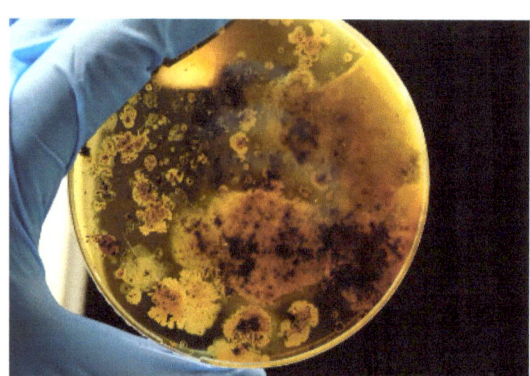

Biological weapons are designed to spread disease among people, plants, and animals through the introduction of toxins and microorganisms such as viruses and bacteria. The method through which a biological weapon is deployed depends on the agent itself, its preparation, its durability, and the route of infection. Attackers may disperse these agents through aerosols or food and water supplies (1).

Although bioweapons have been used in war for many centuries, a recent surge in genetic understanding, as well as a rapid growth in computational power, has allowed genetic engineering to play a larger role in the development of new bioweapons.

In the bioweapon industry, genetic engineering can be used to manipulate genes to create new pathogenic characteristics aimed at enhancing the efficacy of the weapon through increased survivability, infectivity, virulence, and drug resistance.

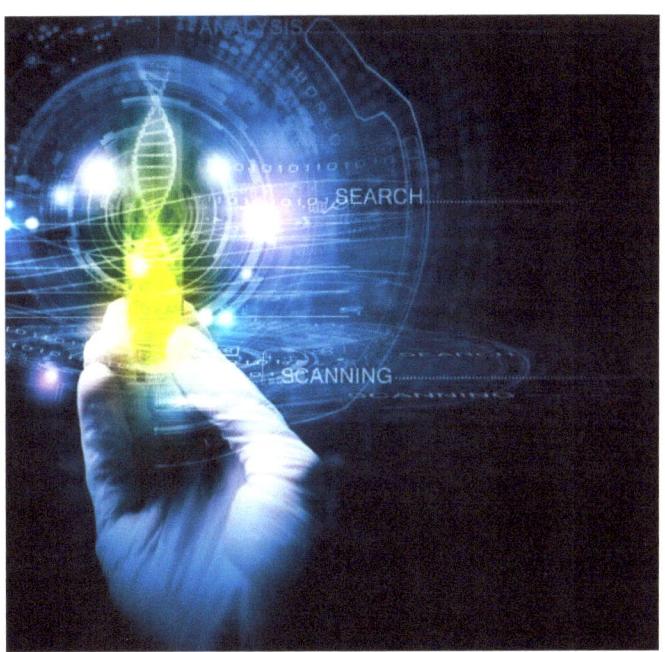

While the positive societal implications of improved biotechnology are apparent, the "black biology" of bioweapon development may be "one of the gravest threats we will face"

Limits of Past Bioweapons

Prior to recent advances in genetic engineering, bioweapons were exclusively natural pathogens. Agents must fulfill numerous prerequisites to be considered effective military bioweapons, and most naturally occurring pathogens are ill suited for this purpose.

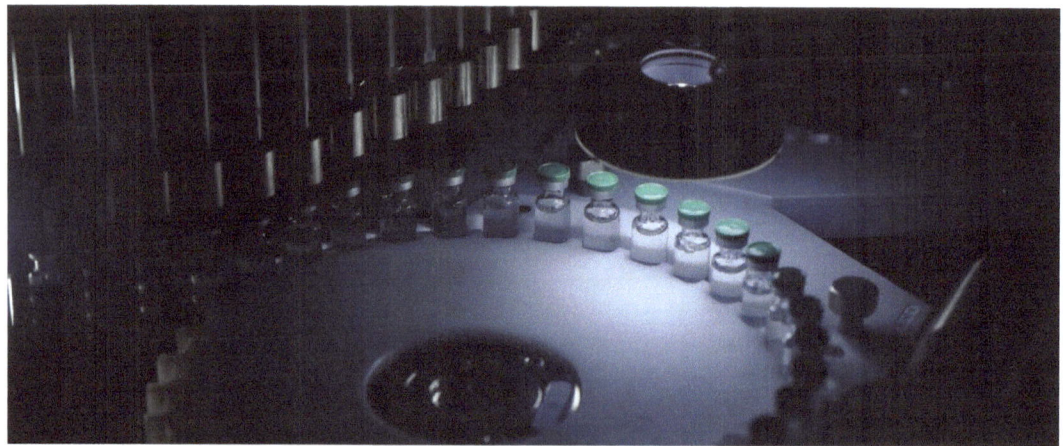

First, bioweapons must be produced in large quantities. A pathogen can be obtained from the natural environment if enough can be collected to allow purification and testing of its properties. Otherwise, pathogens could be produced in a microbiology laboratory or bank, a process which is limited by pathogen accessibility and the safety with which the pathogens can be handled in facilities.

To replicate viruses and some bacteria, living cells are required. The growth of large quantities of an agent can be limited by equipment, space, and the health risks associated with the handling of hazardous germs. In addition to large-scale production, effective bioweapons must act quickly, be environmentally robust, and their effects must be treatable for those who are implementing the bioweapon.

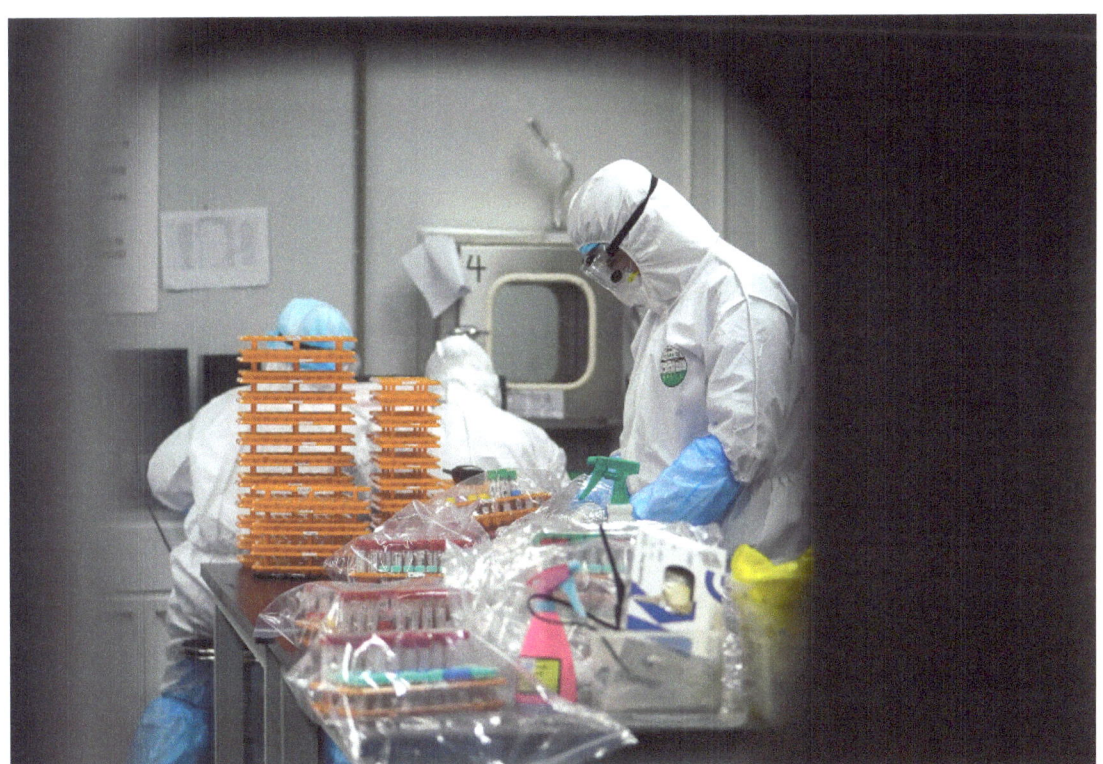

Recent Advances

As researchers continue to transition from the era of DNA sequencing into the era of DNA synthesis, it may soon become feasible to synthesize any virus whose DNA sequence is known. This was first demonstrated in 2001 when Dr. Eckard Wimmer re-created the poliovirus and again in 2005 when Dr. Jeffrey Taubenberger and Terrence Tumpey re-created the 1918 influenza virus. The progress of DNA synthesis technology will also allow for the creation of novel pathogens. According to biological warfare expert Dr. Steven Block, genetically engineered pathogens "could be made safer to handle, easier to distribute, capable of ethnic specificity, or be made to cause higher mortality rates".

The growing accessibility of DNA synthesis capabilities, computational power, and information means that a growing number of people will have the capacity to produce bioweapons. Scientists have been able to transform the four letters of DNA—A (adenine), C (cytosine), G (guanine), and T (thymine)—into the ones and zeroes of binary code.

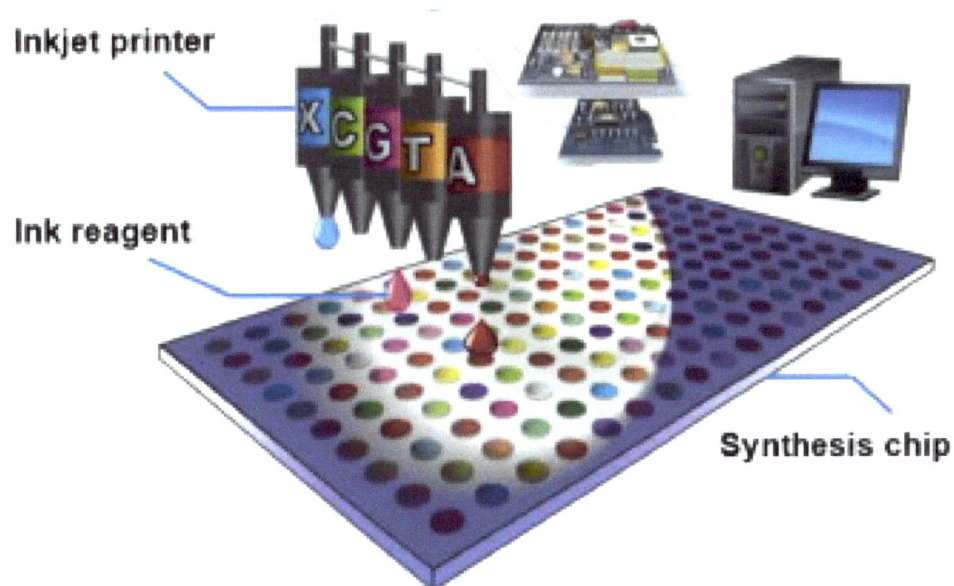

This transformation makes genetic engineering a matter of electronic manipulation, which decreases the cost of the technique. According to former Secretary of State Hillary Clinton, "the emerging gene synthesis industry is making genetic material more widely available [...] A crude but effective terrorist weapon can be made using a small sample of any number of widely available pathogens, inexpensive equipment, and college-level chemistry and biology."

Techniques to Enhance Efficacy of Bioweapons

Scientists and genetic engineers are considering several techniques to increase the efficacy of pathogens in warfare.

1. Binary Biological Weapons

This technique involves inserting plasmids, small bacterial DNA fragments, into the DNA of other bacteria in order to increase virulence or other pathogenic properties within the host bacteria.

2. Designer Genes

According to the European Bioinformatics Institute, as of December 2012, scientists had sequenced the genomes of 3139 viruses, 1016 plasmids, and 2167 bacteria, some of which are published on the internet and are therefore accessible to the public. With complete genomes available and the aforementioned advances in gene synthesis, scientists will soon be able to design pathogens by creating synthetic genes, synthetic viruses, and possibly entirely new organisms.

3. Gene Therapy

Gene therapy involves repairing or replacing a gene of an organism, permanently changing its genetic composition. By replacing existing genes with harmful genes, this technique can be used to manufacture bioweapons.

4. Stealth Viruses

Stealth viruses are viral infections that enter cells and remain dormant for an extended amount of time until triggered externally to cause disease. In the context of warfare, these viruses could be spread to a large population, and activation could either be delayed or used as a threat for blackmail.

5. Host-Swapping Diseases

Much like the naturally occurring West Nile and Ebola viruses, animal viruses could potentially be genetically modified and developed to infect humans as a potent biowarfare tactic.

6. Designer Diseases

Biotechnology may be used to manipulate cellular mechanisms to cause disease. For example, an agent could be designed to induce cells to multiply uncontrollably, as in cancer, or to initiate apoptosis, programmed cell death.

7. Personalized Bioweapons

In coming years it may be conceivable to design a pathogen that targets a specific person's genome. This agent may spread through populations showing minimal or no symptoms, yet it would be fatal to the intended target.

Biodefense

In addition to creating bioweapons, the emerging tools of genetic knowledge and biological technology may be used as a means of defense against these weapons.

1. Human Genome Literacy

As scientific research continues to reveal the functions of specific genes and how genetic components affect disease in humans, vaccines and drugs can be designed to combat particular pathogens based on analysis of their particular molecular effect on the human cell.

2. Immune System Enhancement

In addition to enabling more effective drug development, human genome literacy allows for a better understanding of the immune system. Thus, genetic engineering can be used to enhance human immune response to pathogens. As an example, Dr. Ken Alibek is conducting cellular rese   eapon anth

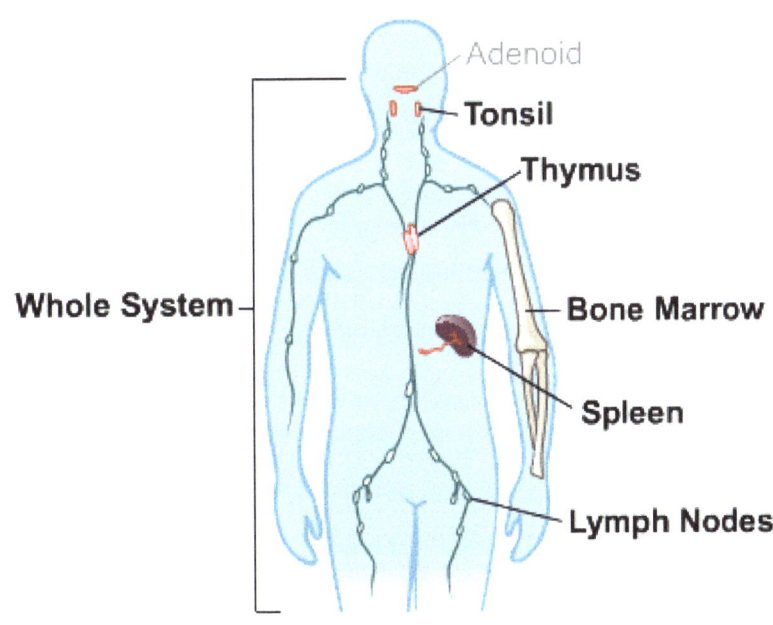

4. Efficient Bio-Agent Detection and Identification Equipment

Because the capability of comparing genomes using DNA assays has already been acquired, such technology may be developed to identify pathogens using information from bacterial and viral genomes. Such a detector could be used to identify the composition of bioweapons based on their genomes, reducing present-day delays in resultant treatment and/or preventive measures.

5. New Vaccines

Current scientific research projects involve genetic manipulation of viruses to create vaccines that provide immunity against multiple diseases with a single treatment.

6. New Antibiotics and Antiviral Drugs

Currently, antibiotic drugs target DNA synthesis, protein synthesis, and cell-wall synthesis processes in bacterial cells. With an increased understanding of microbial genomes, other proteins essential to bacterial viability can be targeted to create new classes of antibiotics. Eventually, broad-spectrum, rather than protein-specific, anti-microbial drugs may be developed.

Future of Warfare

"The revolution in molecular biology and biotechnology can be considered as a potential Revolution of Military Affairs (RMA)," states Colonel Michael Ainscough, MD, MPH (2). According to Andrew Krepinevich, who originally coined the term RMA, "technological advancement, incorporation of this new technology into military systems, military operational advancement, and organizational adaptation in a way that fundamentally alters the character and conduct of conflict" are the four components that make up an RMA.

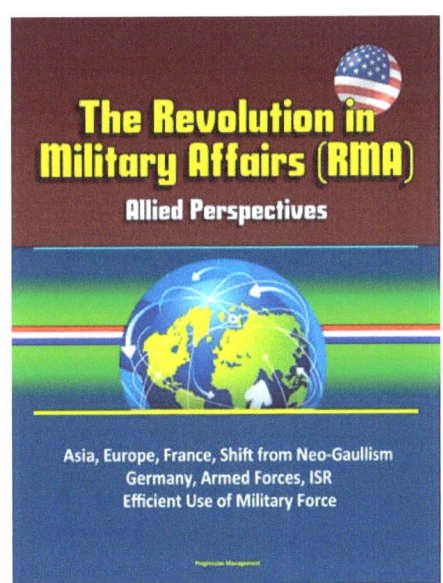

For instance, the Gulf War has been classified as the beginning of the space information warfare RMA. "From the technological advances in biotechnology, biowarfare with genetically engineered pathogens may constitute a future such RMA," says Ainscough.

In addition, the exponential increase in computational power combined with the accessibility of genetic information and biological tools to the general public and lack of governmental regulation raise concerns about the threat of biowarfare arising from outside the military.

The US government has cited the efforts of terrorist networks, such as al Qaida, to recruit scientists capable of creating bioweapons as a national security concern and "has urged countries to be more open about their efforts to clamp down on the threat of bioweapons".

Despite these efforts, biological research that can potentially lead to bioweapon development is "far more international, far more spread out, and far more diverse than nuclear science [...] researchers communicate much more rapidly with one another by means that no government can control [...] this was not true in the nuclear era," according to David Kay, former chief U.S. weapons inspector in Iraq. Kay is "extraordinarily pessimistic that we [the United States] will take any of the necessary steps to avoid the threat of bioweapons absent their first actual use".

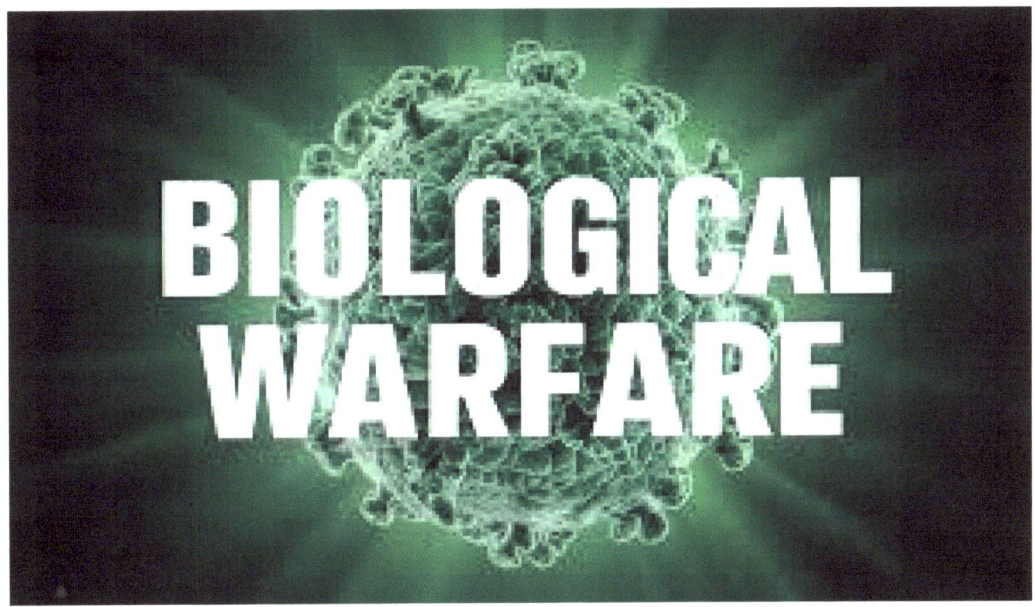

"There are those who say: 'the First World War was chemical; the Second World War was nuclear; and that the Third World War – God forbid – will be biological'".

A common mistake is labeling sGBW as 'Ethnic' bioweapons, while the proper term is 'Race' when referring to biological organisms genetically targeting a population. In light of this clarification, we cite the following introductory article authored by Jim Geraghty, dated April 10, 2023, in the *National Review:*

The Coming Threat of a Genetically-Engineered 'Ethnic Bioweapon'

Excerpts:

Paul Dabbar, undersecretary of energy for science during the Trump administration, writing in the *Wall Street Journal* last week:

Around 2017, the Energy Department's national laboratories started having significant concerns about biosecurity with regard to China. A Chinese general who was head of the National Defense University in Beijing publicly declared an interest in using gene sequencing and editing to develop pathogenic bioweapons that would target specific ethnic groups, which may be the most evil idea I have ever encountered. Taking note, the Commerce Department ordered export restrictions of potentially dangerous biotechnology to China. But the NIH and NIAID refused to believe that there was any risk involved in collaborating with Chinese labs. Their indiscriminate commitment to open science blinds them to threats, even when a country like China is open about its intentions.

Michael Knutzen, a biosecurity specialist who is former Army intelligence and a Presidential Management Fellow at the Department of Homeland Security, wrote at the U.S. Naval Institute:

Excerpts:

Some researchers (including Lieutenant General Zhang Shibo, former president of the PLA National Defense University) foresee the possibility of "specific ethnic genetic attacks" on whole racial or ethnic groups, although there remain political and scientific obstacles at present.

A unique person with unique genes is easier to target than population-level differences in the nearer term. SBWs (Ed. Note: Synthetic Bioweapons) with high levels of asymptomatic transmission could pass from host-to-host through the human domain, until reaching a vulnerable target or targets possessing the "right" genes. Procuring a president or admiral's DNA is easy. Simply invite the target to dinner at a venue you control.

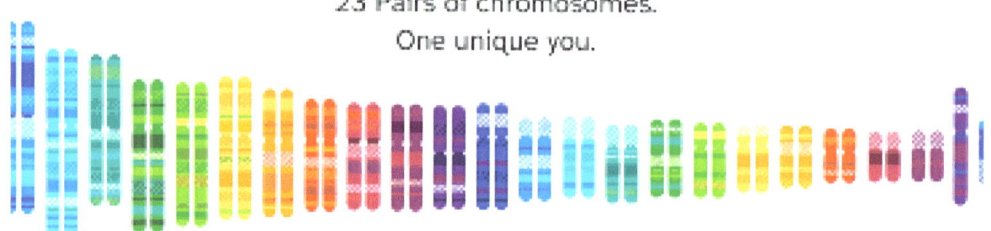

23 Pairs of chromosomes.
One unique you.

And China may already have hacked from medical records or purchased the genetic information of millions of ordinary Americans through genealogical companies such as 23andMe. Bill Evanina, former director of the National Counterintelligence and Security Center, warned against Beijing Genomics Institute–linked COVID-19 tests, noting: "Foreign powers can collect, store and exploit biometric information from COVID tests."

Six Emerging Threats

- Binary Biological Weapons
- Designer Diseases
- Designer Genes
- Gene Therapy as a Weapon
- Stealth Viruses
- Host-swapping Diseases

In addition to the threats posed by sGBW (synthetic Genetic Biological Weapons), are the concepts of somehow controlling, or *'commanding'* biotechnology, to include so-called *'merciful conquest'* of enemy forces. Authored in November 2006 by Ji-Wei Guo in the journal of *Military Medicine*; *The Command of Biotechnology And Merciful Conquest In Military Opposition*, Guo cites:

Excerpts:

"Judging from the evolving law of the theory of command, the command of biotechnology is feasible and inevitable".

"This theory is expected to achieve successes in wars in an ultramicro, nonlethal, reversible, and merciful way and will play an important role in biotechnological identification and orientation, defense and attack, and the maintenance of fighting powers and biological monitoring".

Single Nucleotide Polymorphism

- Single nucleotide polymorphism (SNP) refers to a single base change in a DNA sequence

- SNP: Commonly biallelic

- Two types(Based on presence in genome)
  - ➤Synonymus
  - ➤Non-synonymus

- SNPs have largely replaced simple sequence repeats (SSRs)

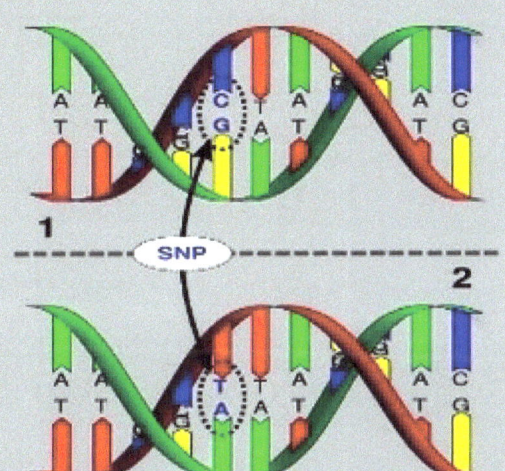

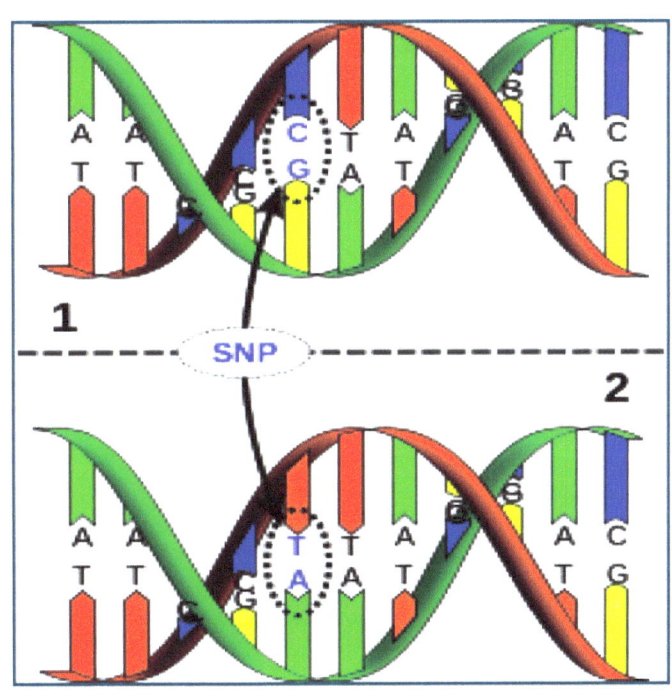

Precision Medicine Targeting Single Nucleotide Polymorphisms

Excerpts:

Single nucleotide polymorphisms, frequently called SNPs (pronounced "snips"), are the most common type of genetic variation among people. Each SNP represents a difference in a single DNA building block, called a nucleotide. For example, a SNP may replace the nucleotide cytosine (C) with the nucleotide thymine (T) in a certain stretch of DNA.

Most commonly, SNPs are found in the DNA between genes._They can act as biological markers, helping scientists locate genes that are associated with disease. When SNPs occur within a gene or in a regulatory region near a gene, they may play a more direct role in disease by affecting the gene's function.

SNPs help predict an individual's response to certain drugs, susceptibility to environmental factors such as toxins, and risk of developing diseases. SNPs can also be used to track the inheritance of disease-associated genetic variants within families. Research is ongoing to identify SNPs associated with complex diseases such as heart disease, diabetes, and cancer.

Scientific Risk Assessment of Genetic Weapon Systems

James Martin Center for Nonproliferation Studies
Middlebury Institute of International Studies at Monterey

Richard Pilch Jill Luster Miles Pomper Robert Shaw

CNS OCCASIONAL PAPER #52 SEPTEMBER 2021

Excerpts

The goal of precision medicine is to identify subpopulations of patients with unique disease susceptibilities, prognoses, and therapeutic responses in order to optimize their medical management, from basic preventive measures such as preemptive lifestyle modifications to advanced treatments.

In clinical practice, precision medicine considers genetic, lifestyle, and environmental factors, but from a pharmaceutical industry perspective the focus is on identifying common genetic factors that can be "precisely" targeted by medical countermeasures. Scientific research has long indicated that such common genetic factors can lead to different drug responses across regional, ethnic, and racial subpopulations.

SNPs inform precision medicine approaches employing a technique called "pharmacogenomics." Pharmocogenomics is the study of how genomic differences affect pharmaceutical responses across different subpopulations, in terms of both effectiveness and adverse effects.

If a SNP alters gene expression in a given drug's absorption, distribution, metabolism and excretion (ADME) pathway, then that drug may be more or less effective, or dangerous in populations with the SNP.

SINGLE NUCLEOTIDE POLYMORPHISM

- Single nucleotide polymorphisms or SNP (pronounced "snips"), are the most common type of genetic variation among peoples.

- Each SNP represents a difference in a single DNA building block, called a nucleotide

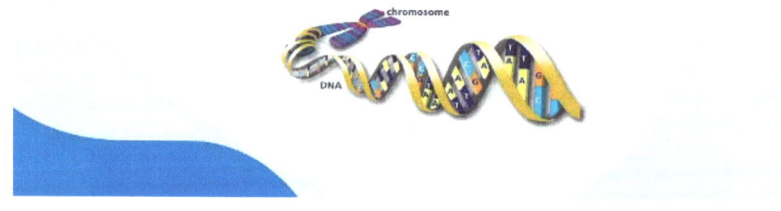

Such relationships can be validated using real-world data of populations containing a SNP or SNPs that map to a given drug's ADME pathway exhibit different responses (efficacy or adverse effects) to that drug.

Various pharmacogenomic methodologies for identifying SNPs, determining which SNPs are drug-related, and predicting corresponding drug effects in different regional, ethnic, and racial subpopulations (termed "pharmacoethnicity") have been described in the open source domain. The resulting SNP-drug effect relationships are curated in a number of openly accessible databases,39 which will continue to grow as efforts to map an increasing number of human genomes come to fruition.

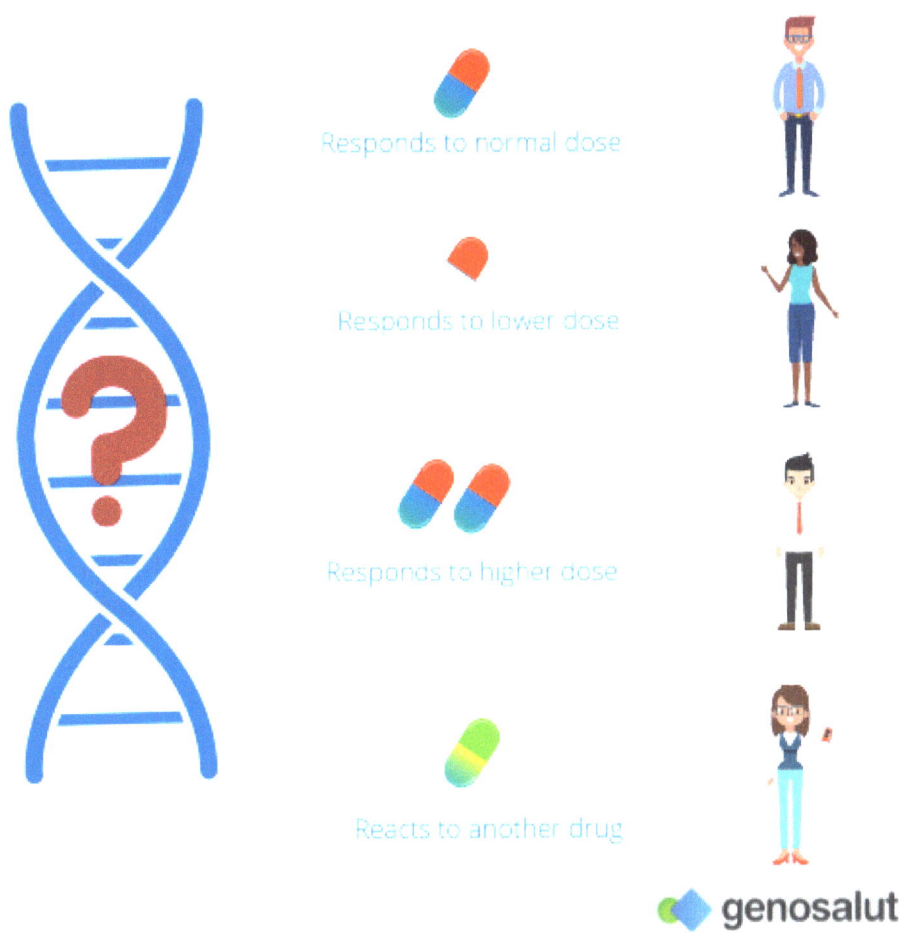

Synthetic Genetic Biological Weapons (sGBW) can also be referred to as "Black" biology, defined as:

> *"The use of genetic engineering to enhance the virulence of a pathogen or the targeting of a specific genetic code for us in terrorism.This new area of biology could create a designer virus which, while initially mimicking the common cold or flu, could act as a—molecular key to trigger secondary effects after encountering a certain DNA sequence."* - *NICOLE H. KALUPA, Wisconsin International Law Journal 5/15/2017*

BLACK BIOLOGY: GENETIC ENGINEERING, THE FUTURE OF BIOTERRORISM, AND THE NEED FOR GREATER INTERNATIONAL AND COMMUNITY REGULATION OF SYNTHETIC BIOLOGY

NICOLE H. KALUPA

Wisconsin International Law Journal 5/15/2017

Excerpts

Synthetic biology, or genetic engineering, is the integration of a multitude of scientific disciplines that seek to alter human DNA at a fundamental level. With this technology, scientists are able to edit out undesirable sequences for the benefit of human health or create new biological material from DNA building blocks. However, this new technology has a black side.

This type of modification could prove useful if the secondary effects delivered individualized cancer treatments to afflicted patients, ensuring that their bodies accepted treatment. But more sinisterly, a virus could be designed with secondary effects inducing the neurodegenerative, fatal byproducts of botulinum toxin and the DNA sequence engineered for recognition could be that of the President of the United States.

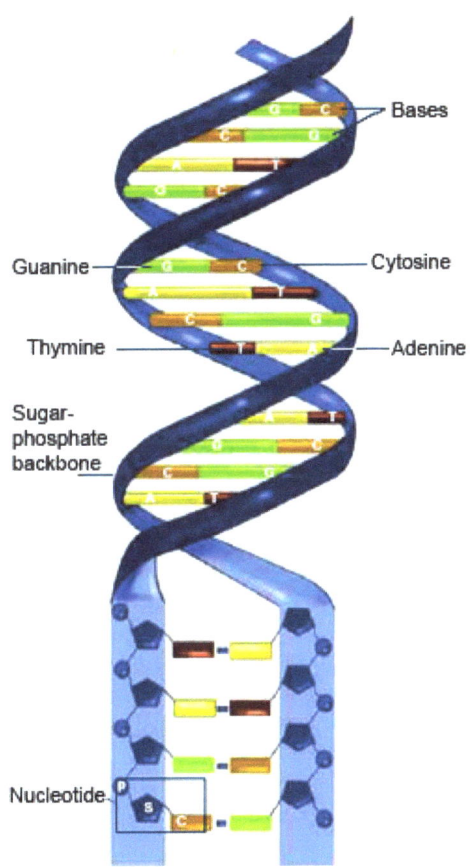

Capitalizing on new developments in computer technology and more readily available genetic material, synthetic biologists attempt to create artificial life and reverse-engineer the building blocks of humanity.

In extremely simplistic terms, synthetic biology is genetic engineering. Scientists utilize bioinformatics to reconstruct the proteins and enzymes encoded in the DNA sequence. This process functions by signaling and manipulating the pathways to produce biological functions that are corrected or more desirable than their original functions. Whether genetic editing/engineering is done to the entire genome or a singular gene, it produces staggering and sometimes unpredictable effects. Most of the modeling is done using computational technologies and other bioinformatics techniques to predict the effects certain manipulations will have on human cells.

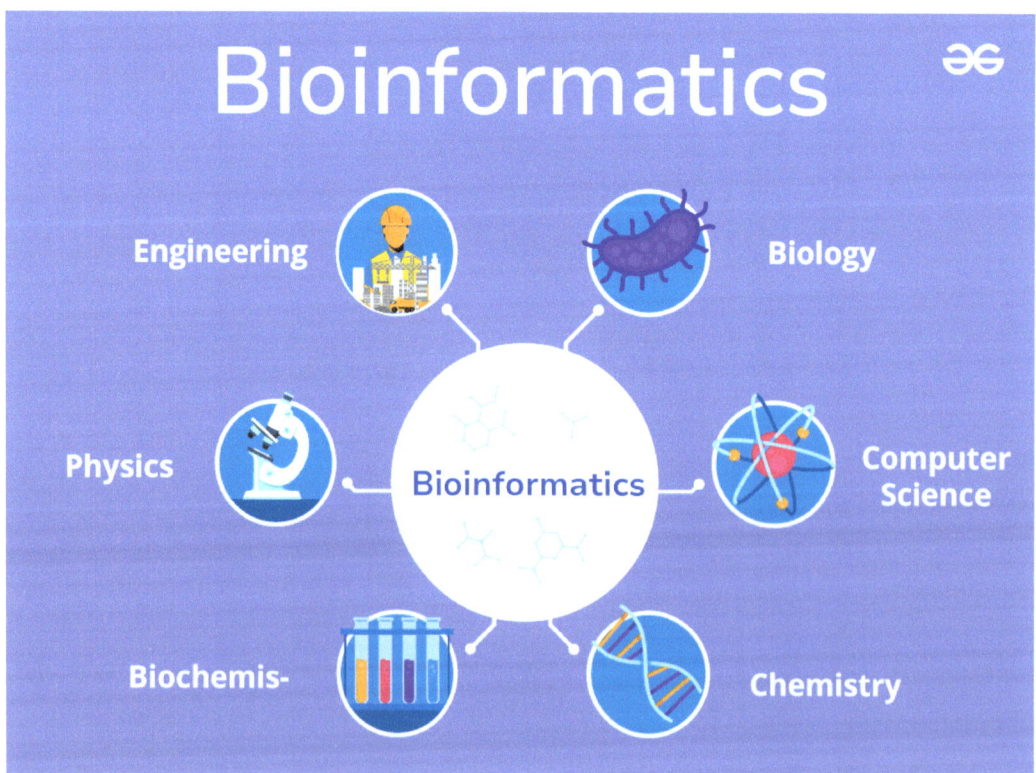

Scientists have identified three categories of risk incited by developments in synthetic biology:

First, —synthetic microorganisms might escape from a research laboratory or containment facility, proliferate out of control and cause environmental damage or threaten public health. This scenario has already played out in traditional manifestations of state-sponsored bioweapon research development.

Second,—a synthetic microorganism developed for some applied purpose might cause harmful side effects after being deliberately released in to the open environment.

Third,—outlaw states, terrorist organizations or individuals might exploit synthetic biology for hostile or malicious purposes.

Creating a *de novo* DNA synthesis is rapidly becoming easier through the use of DNA synthesizers. These machines allow researchers to assemble novel and existing genetic sequences using readily accessible reagents.

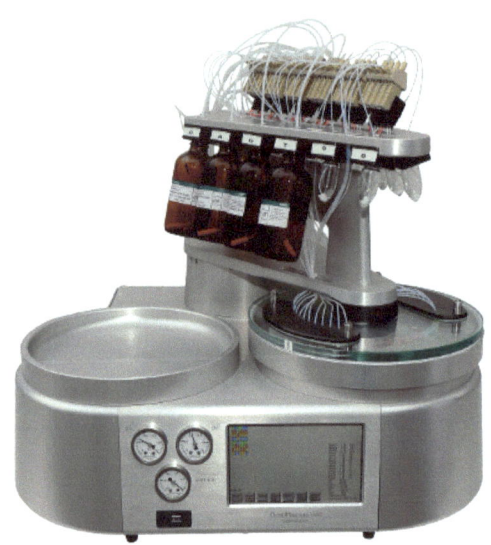

The most simplistic way to construct a genetic sequence is to order a gene or genome-length stretch of viral or bacterial DNA from a commercial gene synthesis company. There are currently forty-five organizations worldwide that have this capacity, with twenty-four companies located in the United States.

After obtaining DNA, an individual could utilize it for the purposes of synthetic biology and endeavor to make modifications that would increase the pathogenicity of the organism. These purchases are closely tracked, particularly in the United States, where especially potent viral DNA strands such as anthrax and others are monitored by the US government.

Alternatively, a researcher could start with smaller pieces of DNA called oligonucleotides or oligos. Oligos are DNA building blocks of 15-100 base pairs that can be linked together to construct gene and genomic length DNA sequences. As oligos are commercially available, this process is understandably more difficult to monitor. From these two options, motivated individuals can replicate bacteria and viruses for their personal and potentially reprehensible research.

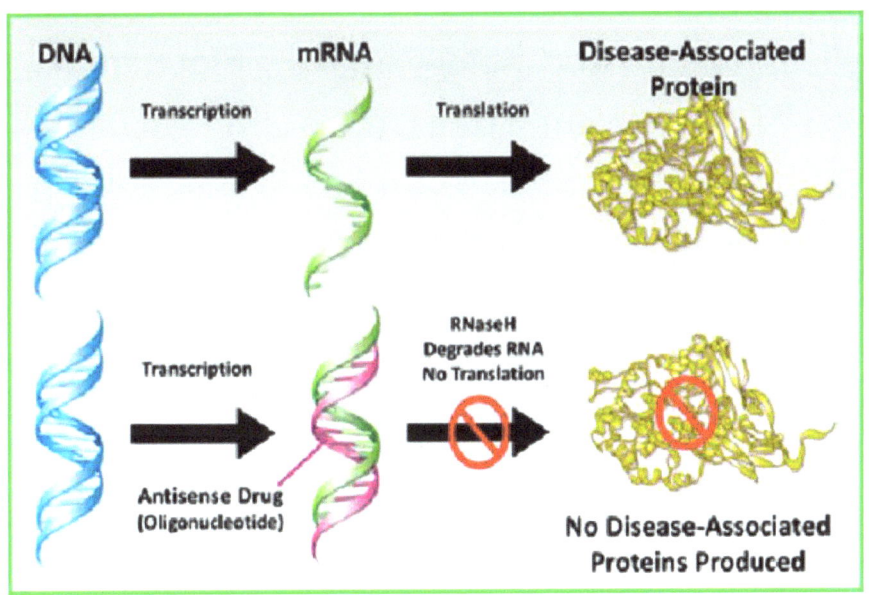

Besides constructing a novel virus genome from scratch, other methods are available to bioterrorists. Currently, replicating genomes requires advanced technology and knowledge. While the goal of synthetic biologists is to make the replication process cheaper and easier to access for research purposes, it would still be difficult, though not impossible, for a non-state sponsored organization or terrorist group to utilize this process.

The Command Of Biotechnology And Merciful Conquest In Military Opposition

Guo JW.

J Spec Oper Med. 2009 Winter; 9(1):69-73.
doi: 10.55460/PWZR-55N0. PMID: 19813351.

Excerpts

The meanings of the command theory of biotechnology consist of: taking a whole or partial lead in the military application of biotechnology; making biotechnology a real power of defense and attack; maintaining a long-lasting advantage in competition of military biotechnology on a large scale.

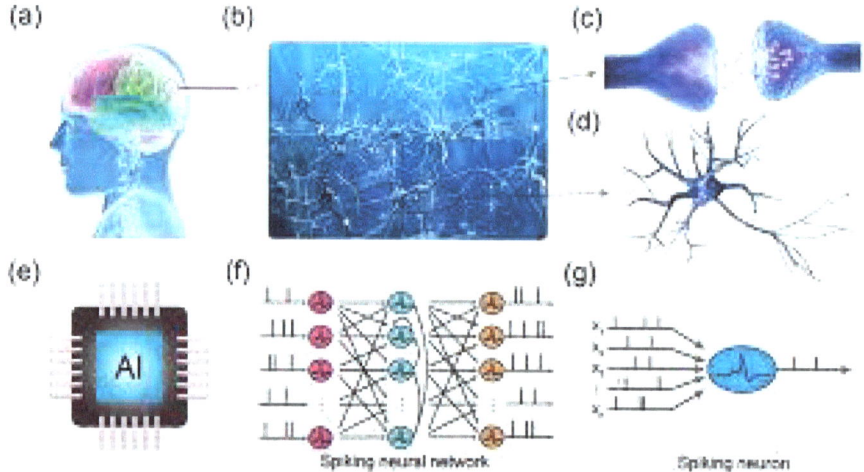

The development of modern biotechnology makes it possible to set up new generation command systems by using biocomputing, sensors, or simulated detectors, which greatly elevates the level of the information-based command platform.

Affecting the structure and function of a gene or a protein as a damaging effect can cause human physiological dysfunction. Precision injury and ultramicro damage are two wounding methods of modern biotechnologies based on genomics and proteomics. They are completely different from the traditional wars that damage tissues and organs directly since they target the primary structure of gene or protein.

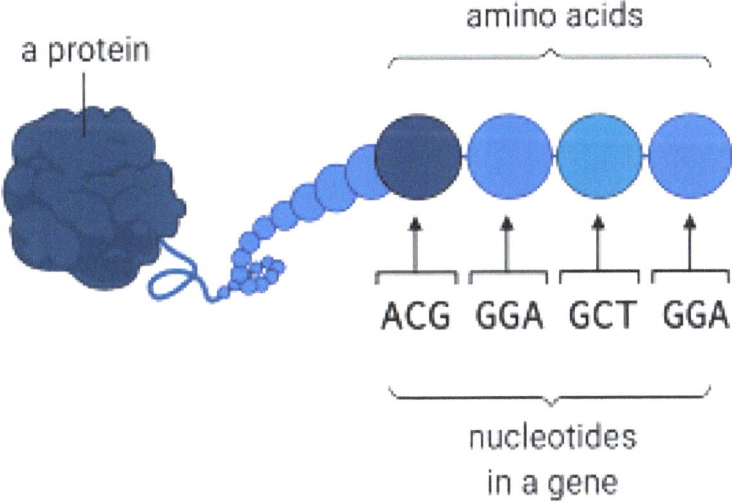

The injuries are completely different from those caused by traditional weapons, including nuclear and chemical weapons. Traditional weapons aim at killing and demolishing in an extreme way. The goal of precision injury is not necessarily to terminate a life, but to choose a degree of injury depending on the purposes of operations and the types of enemies. By means of gene regulation, certain, or a couple of, key physiological functions in a human body - such as learning, memorizing, balancing, fine manipulation, and even the "bellicose" character - can be injured precisely without a threat of life.

The establishment and enrichment of biological informatics embodies the rapid development of modern biotechnology, which is concerned with genes and sequences, structures, and functions of proteins that reveal the mysteries of life. The scale of the top three databases of biological informatics is expanding by geometric series.

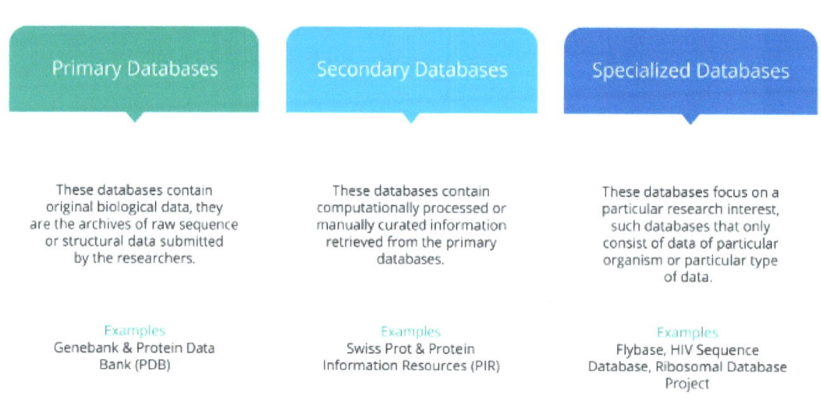

Types Of Biological Databases

Primary Databases

These databases contain original biological data, they are the archives of raw sequence or structural data submitted by the researchers.

Examples
Genebank & Protein Data Bank (PDB)

Secondary Databases

These databases contain computationally processed or manually curated information retrieved from the primary databases.

Examples
Swiss Prot & Protein Information Resources (PIR)

Specialized Databases

These databases focus on a particular research interest, such databases that only consist of data of particular organism or particular type of data.

Examples
Flybase, HIV Sequence Database, Ribosomal Database Project

The development of modern biotechnology is also embodied by the innovation and perfection of many biological techniques and methods, including DNA recombination, gene modification, gene cloning, exogenous gene expression synergy, gene targeting, stem cell technology, and tissue engineering, etc. These technical tools have greatly promoted understanding of life and helped to clarify the relationship between life pattern and military struggle for humans. Therefore, they are possibly to be applied to military purposes.

The invention of the electron microscope makes it possible for us to observe a life structure less than 1 angstrom. What is more, our exploration for the nature of life has reached the molecular level of a protein or a gene. Now that the military the ory of command is to conquer a certain space in battles, either the land, sea, air, or outer space, if technical conditions permitted, the cognitive extension of human beings into the ultramicro space is reasonable and inevitable.

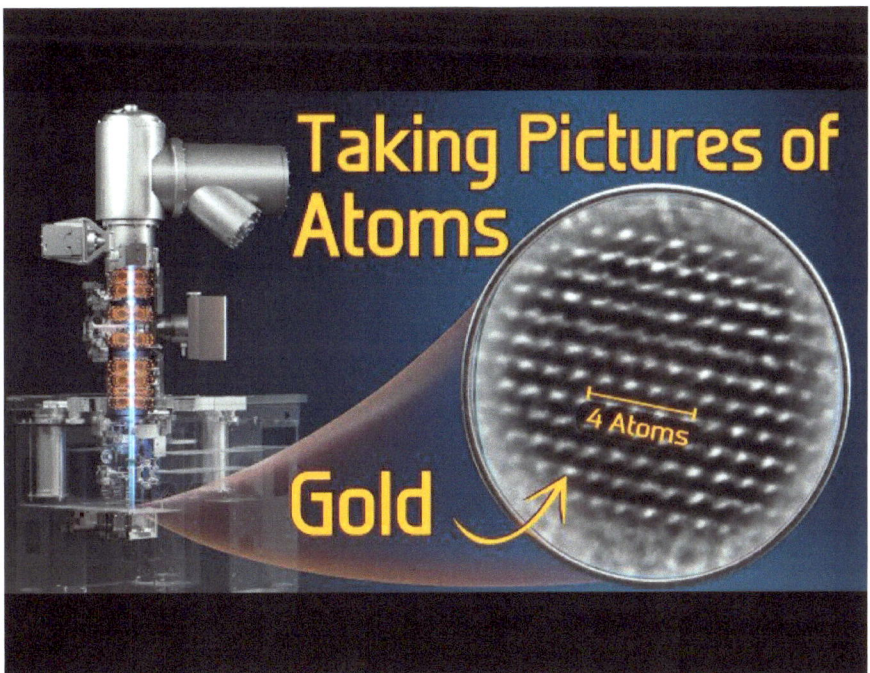

Revolutionary breakthroughs on biotechnology have been made by the progress of science and technology. It has not only brought a more accurate understanding of life itself, but also the power of regulation. Modern biotechnical development has changed the former attributes of biotechnology in military applications.

In the past, biotechnology was mainly used in the prevention, diagnosis, and treatment of injuries and diseases. Now, discoveries made in the exploration of human health through biotechnological methods can clarify the law of life at the molecular level, which makes it possible to regulate and control the functions of human bodies by adjusting its ultramicro structures to gain powers of defense and attack.

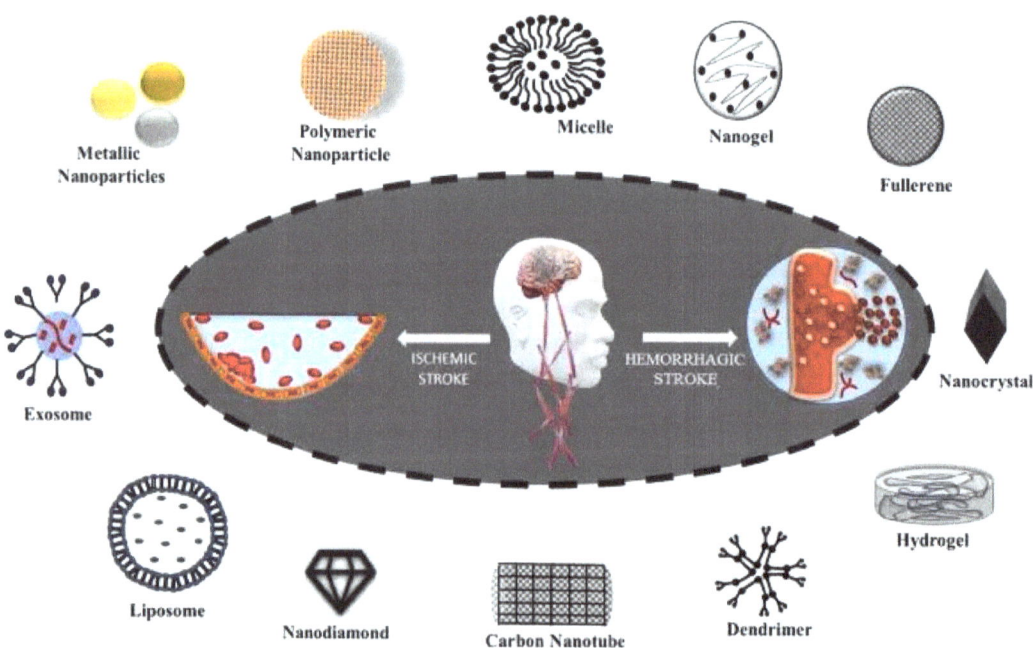

Since war is an act of violence aiming at annihilating enemies or depriving them of resistant abilities, the modern biological techniques used for attack purposes have a more direct and precise target at humans than other methods, which will play a more important role in future military operations.

It bears motivation to pursue social benefits and has a wide developmental prospect. In the last decade, the international productive value of the biotechnological industry increased by five times every 3 years. In developed countries, the increasing speed is approximately 25% to 30%. In the 21st century, the scale of industries related to biological economy will be 10 times that of the information technology industry, which will dominate in international economic growth. Therefore, an effort made to lead and control in the biotechnological field not only has military significance, but can also cement our comprehension.

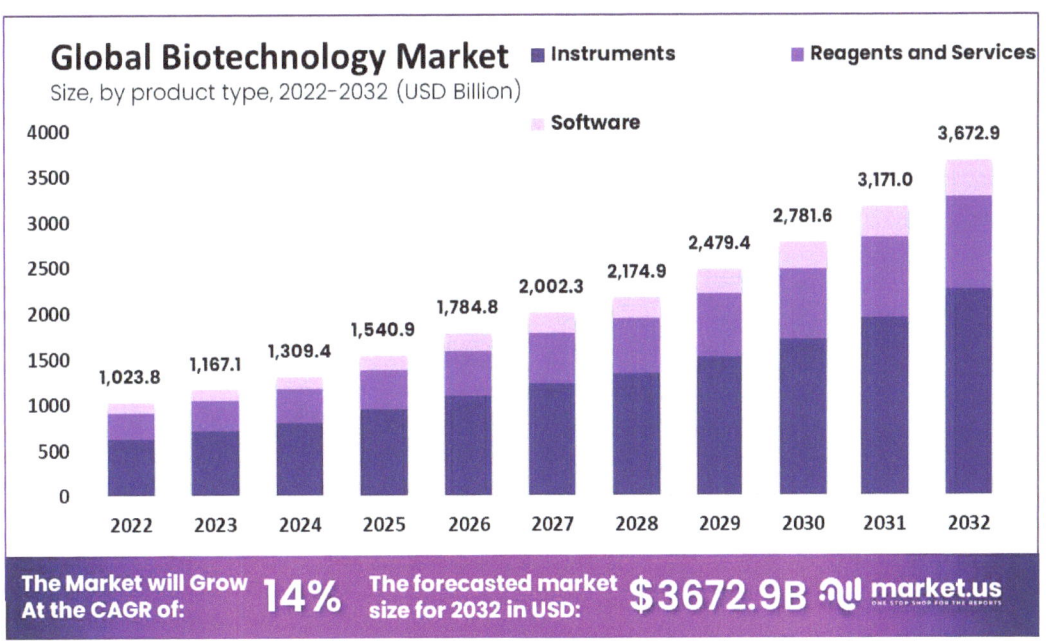

Biotechnology can be used for aggressive purposes, which is the key factor for command of biotechnology. The new categories of injury that may arise are the focus of interest of military medicine.

Modern biotechnology reveals pathologies about factors that do great harm to people and provides effective means of exploring the hazardous factors in human health. Meanwhile, the knowledge can be used to bring damages and injuries to individuals in war in a more accurate and effective fashion. Different military biotechnologies can be chosen in accordance with different pathogenic factors to meet different military goals.

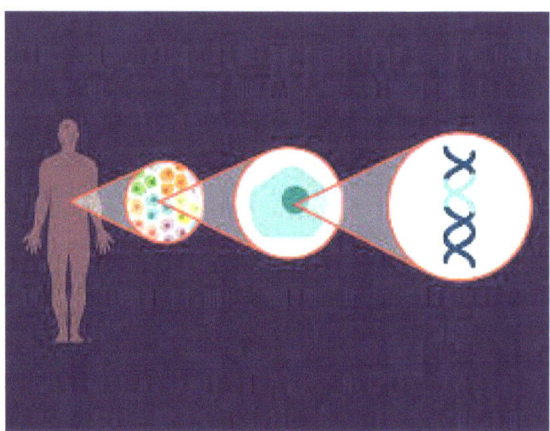

The attack, therefore, will wound different levels of specific gene, protein, cell, tissue, and organ. It no doubt will be more effective to cause damages than conventional weapons, yet the nonlethal effect will remain to be civilized in terms of postwar reconstruction and hatred control.

With ultrastructural damage, targets are chosen directly from a nucleotide sequence or a certain protein structure. Affecting the structure and function of a gene or a protein as a damaging effect can cause human physiological dysfunction. Precision injury and ultramicro damage are two wounding methods of modern biotechnologies based on genomics and proteomics. They are completely different from the traditional wars that damage tissues and organs directly since they target the primary structure of gene or protein.

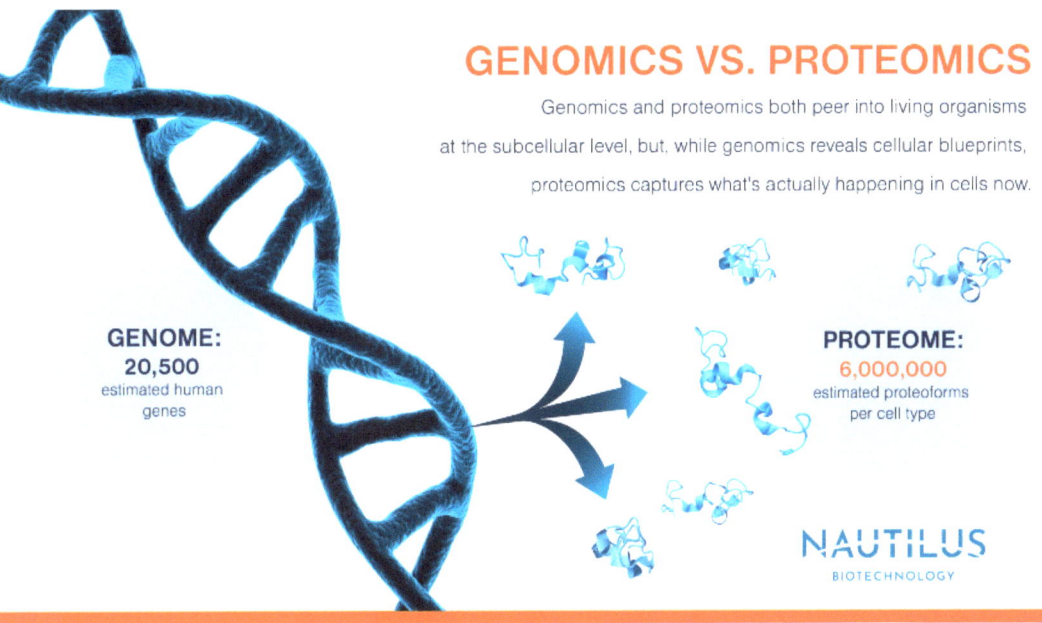

GENOMICS VS. PROTEOMICS

Genomics and proteomics both peer into living organisms at the subcellular level, but, while genomics reveals cellular blueprints, proteomics captures what's actually happening in cells now.

GENOME:
20,500
estimated human genes

PROTEOME:
6,000,000
estimated proteoforms per cell type

NAUTILUS
BIOTECHNOLOGY

The development of modern biotechnology makes it possible to set up new generation command systems by using biocomputing, sensors, or simulated detectors, which greatly elevates the level of the information-based command platform.

Traditional weapons cause body damage, and the effect should be judged on the battlefield. However, the damage of biotechnology can be predicated before war or even in laboratories. Therefore, the damaging capability, targets, and degree of damage can be determined according to the situation, greatly increasing the controllability of war, and realizing fighting effects-based outcomes.

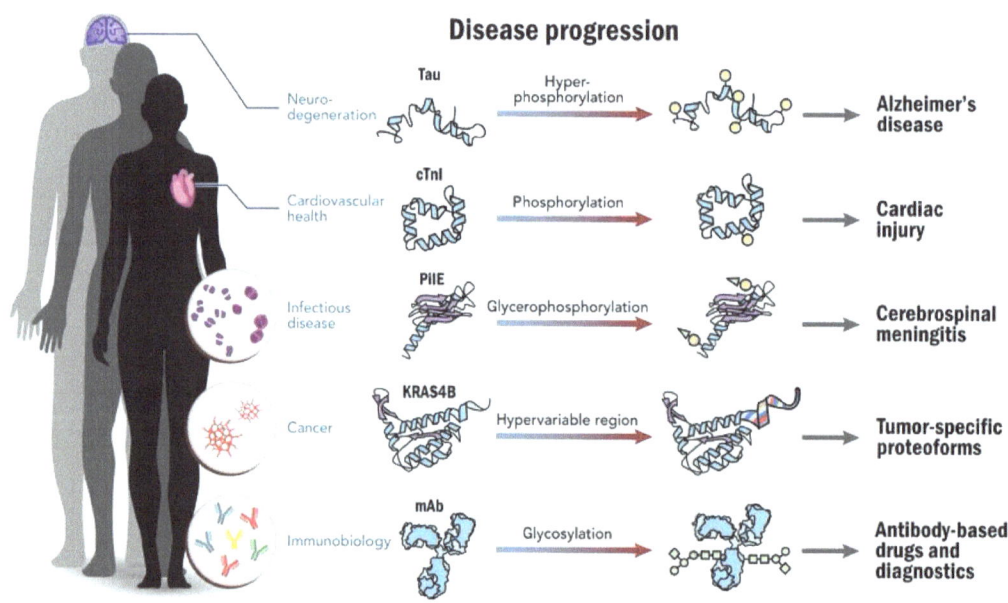

Minds At War: China's Pursuit Of Military Advantage through Cognitive Science and Biotechnology

By Elsa B. Kania

National Defense University Press

PRISM No. 8, Vol. 3

Jan. 9, 2020

Excerpts

In future conflict, the battlefield is expected to extend into new virtual domains. According to He Fuchu, "The sphere of operations will be expanded from the physical domain and the information domain to the domain of consciousness; the human brain will become a new combat space." Consequently, success on the future battlefield will require achieving not only "biological dominance" but also "mental/cognitive dominance" and "intelligence dominance".

At the same time, the notion of "winning without fighting" is a traditional element of Chinese strategic thinking that possesses enduring relevance in an era in which technology is becoming ever more consequential to strategic competition in peacetime.

Chinese military scientists and strategists have often been animated in their thinking by concern with the progression of the ongoing revolution in military affairs (RMA) that is believed to be catalyzed by today's technological advancements. The PLA has closely examined the U.S. military's approach to warfare, applying lessons learned to its own military modernization in seeking to catch up, while also looking for opportunities to pursue asymmetric capabilities or attempt to achieve a first-mover advantage to overtake this "powerful adversary".

Since the 1990s, Chinese military modernization has particularly concentrated on pursuing a strategy of "informatization". Through this agenda, the PLA has developed an array of command, control, communications, computers, intelligence, surveillance, and reconnaissance (C4ISR) systems and concentrated on advancing capabilities for information operations, including cyber warfare, electronic warfare, and psychological warfare.

Today, PLA strategists anticipate a new style of warfare is on the horizon, as the character of conflict evolves from informatized toward "intelligentized" warfare, in which AI, along with a range of technologies, is changing the form of warfare.

According to Lt. Gen. Liu Guozhi, Director of the Central Military Commission Science and Technology Commission, "AI will accelerate the process of military transformation, ultimately leading to a profound Revolution in Military Affairs . . . The combination of artificial intelligence and human intelligence can achieve the optimum, and human-machine hybrid intelligence will be the highest form of future intelligence."

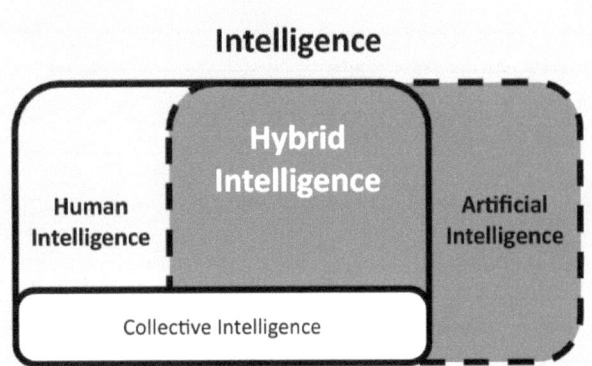

This striking statement highlights the PLA's interest at the highest levels in the notion of "hybrid intelligence", a concept that implies a blending of human and machine intelligence, including through leveraging insights from brain science and such techniques as the use of brain-computer interfaces.

This concept is not merely abstract but is starting to be realized through new programs, including projects intended to promote human performance enhancement. Future intelligentized operations are expected to involve prominent employment of intelligent autonomy in weapons systems under conditions of multi-domain integration with command exercised through brain-machine integration, enabled by cloud infrastructure.

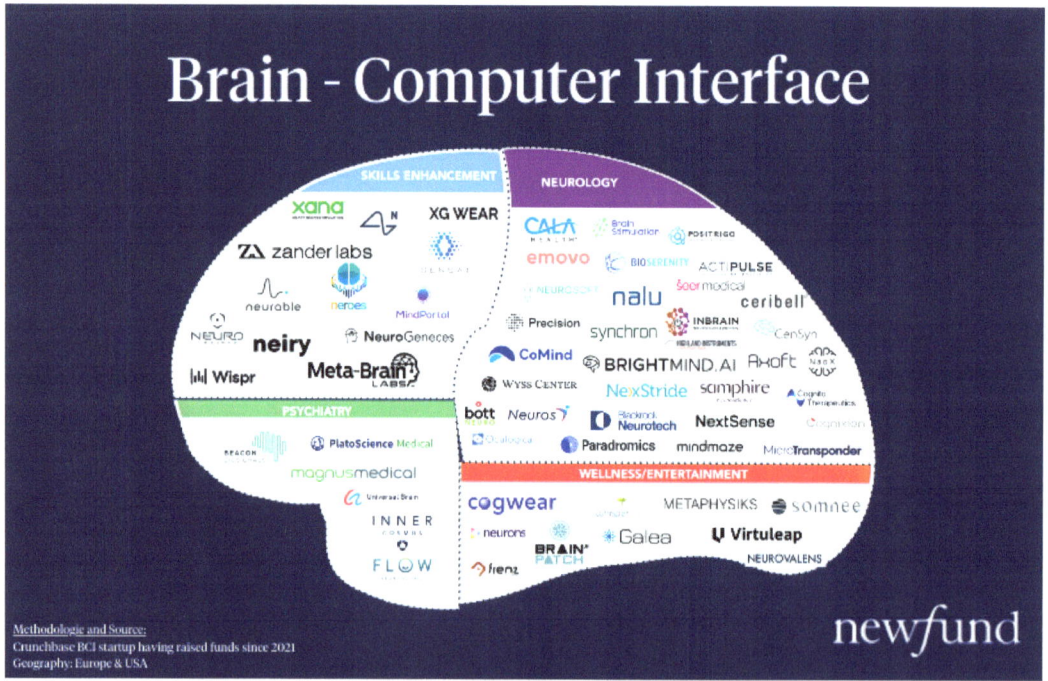

Chinese military scientists and strategists expect that this revolution in warfare will also demand transformation of the human element of warfare, which may require seeking command of the brain and biological sciences.

U.S. Intelligence Claims China Wants to Steal Your DNA

By Lucas Ropek

Gizmodo.com

January 2021

Excerpts

The Chinese biotech firm BGI Group recently offered to "build and run" Covid-19 testing centers in multiple U.S. states—including California, New York, and half a dozen others—but Bill Evanina, one of the top federal intelligence officials in the country, issued a dire warning against it, according to a new report from "60 Minutes."

Evanina, who at the time served as director of the National Counterintelligence and Security Center, sent a bulletin to governments and hospitals warning them that "foreign powers can collect, store and exploit biometric information from covid tests." In a recent CBS interview, Evanina made an even bolder claim: that the reason the Chinese are trying to collect Americans' data is to "win a race to control the world's biodata."

National security concerns about the DNA threat seem to boil down to the idea that if the Chinese have too much data on our genetics it will give them undue influence over us politically. A 2029 report prepared for the U.S.–China Economic and Security Review Commission claims that China might use the DNA it has collected to make targeted attacks on "sensitive US persons,

Indeed, warnings about China's alleged desire to gobble up the world's genetic data and use it for nefarious purposes have been ongoing for some time. It's true that China has an extensive domestic DNA collection program, having launched an initiative to create a national genetics database in 2017. Concerns exist that this data will be used to control trends in medicine and pharmaceuticals, or to engineer bioweapons.

The genomics firm at the center of the most recent drama, BGI, is one of the biggest in the world. The company, which specializes in genome sequencing, substantially increased its operations when Covid-19 struck last year. As the virus got underway in the U.S., an affiliate of the company began "approaching city, county and state officials with offers to sell supplies and help set up entire labs, proposing to export a rapid testing model that they said had helped contain China's outbreak," the Washington Post reports. BGI has denied that it wants to collect American DNA via Covid tests.

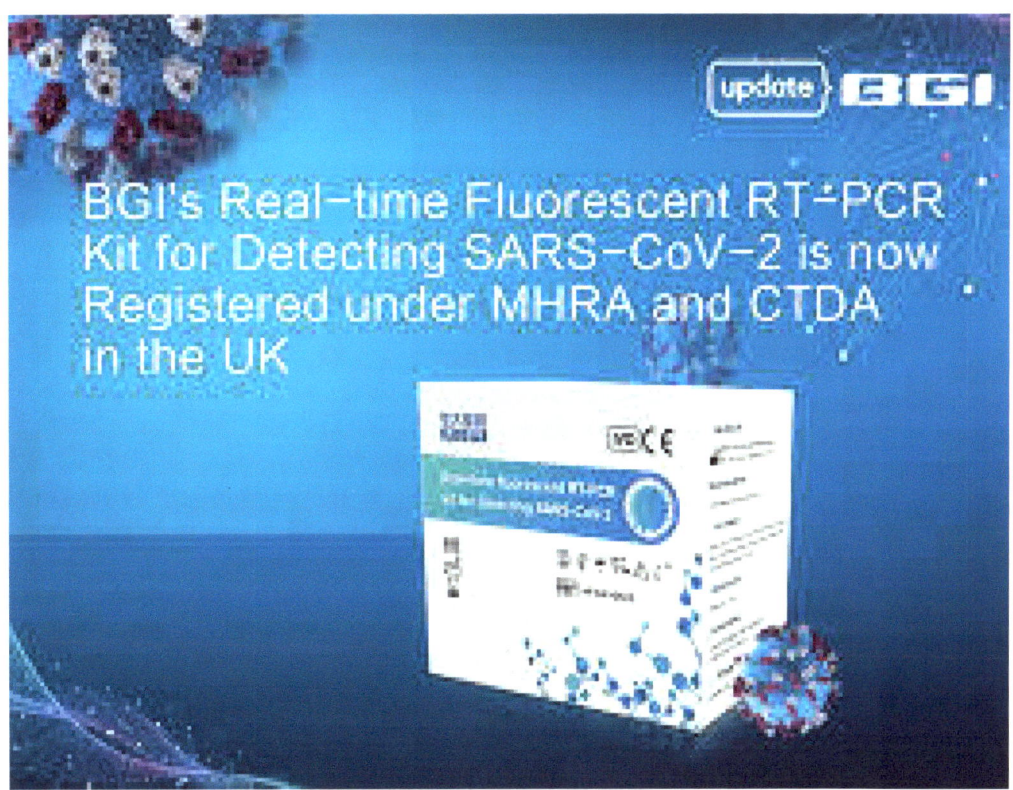

At the same time, it can't help but be noted how China's advances in biotechnology appear to be spurring the U.S. to do the same, as pressures mount for the federal government to invest further in similar DNA collection and research. The same U.S.–China Commission report claims that America already "maintains a superior biotechnology innovation" than China but that "continued investment by the US in its own biotechnology industry will go a long way toward limiting the effectiveness of China's efforts to close the biotechnology gap between the two countries."

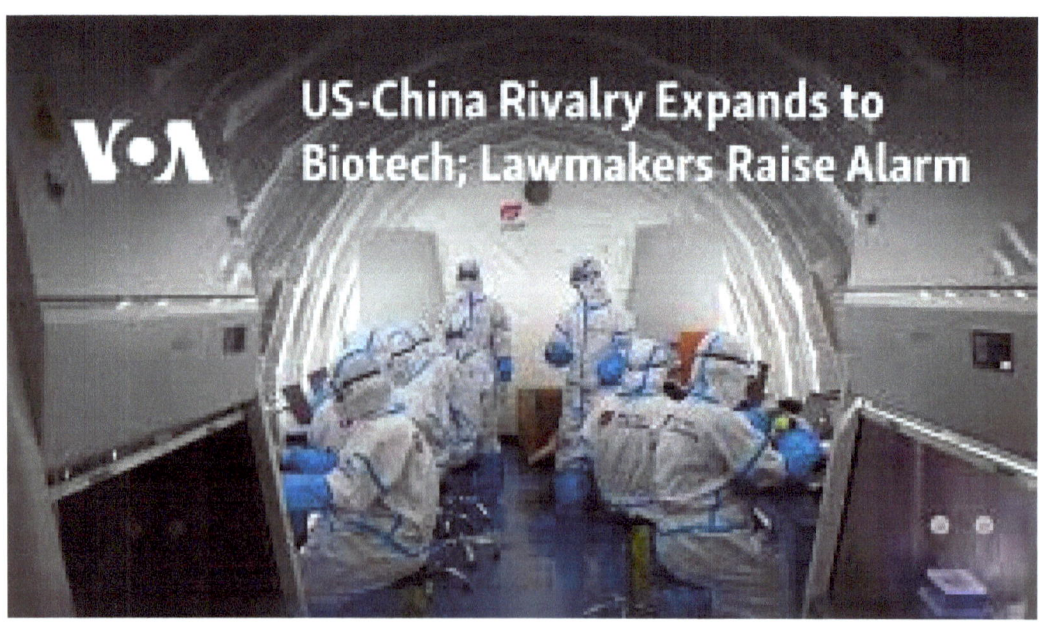

Targeted Genome Modification Via Triple Helix Formation.

Ricciardi AS, McNeer NA, Anandalingam KK,
Saltzman WM, Glazer PM.

Methods Mol Biol. 2014;
1176:89-106. doi: 10.1007/978-1-4939-0992-6_8.
PMID: 25030921; PMCID: PMC5111905.

Abstract

Triplex-forming oligonucleotides (TFOs) are capable of coordinating genome modification in a targeted, site-specific manner, causing mutagenesis or even coordinating homologous recombination events. Here, we describe the use of TFOs such as peptide nucleic acids for targeted genome modification. We discuss this method and its applications and describe protocols for TFO design, delivery, and evaluation of activity in vitro and in vivo.

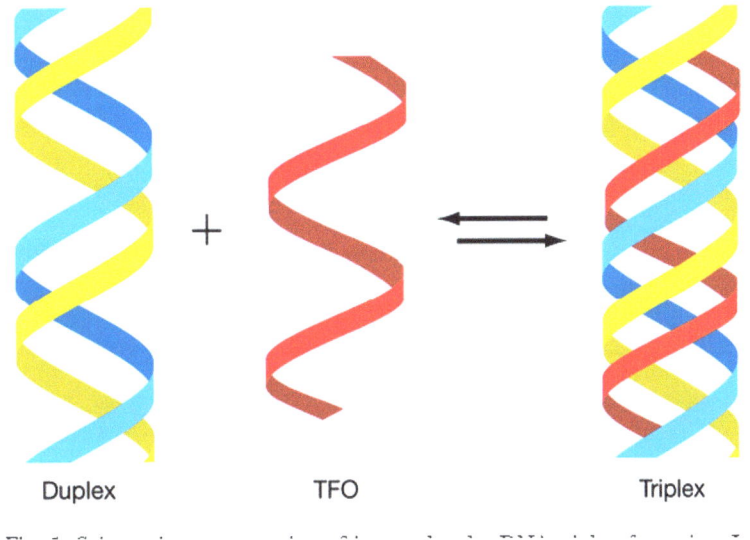

Duplex   TFO   Triplex

Excerpts

While double helices are key to the understanding and study of the biological sciences, nucleic acids are also capable of forming triple helices. In fact, before the establishment of the double-helical nature of DNA, Linus Pauling proposed a triple-helix structure. Felsenfeld et al. demonstrated the possibility of triple helix formation when they noted that polyU and polyA RNA strands could bind in a 2:1 ratio.

Triplex-forming oligonucleotides, or TFOs, can form similar triple helices. TFOs can bind in the major groove of duplex DNA in a polypurine/polypyrimidine run, with reverse Hoogsteen hydrogen bonds antiparallel to a polypurine strand of a DNA duplex or with Hoogsteen bonds in a parallel orientation to the purine strand.

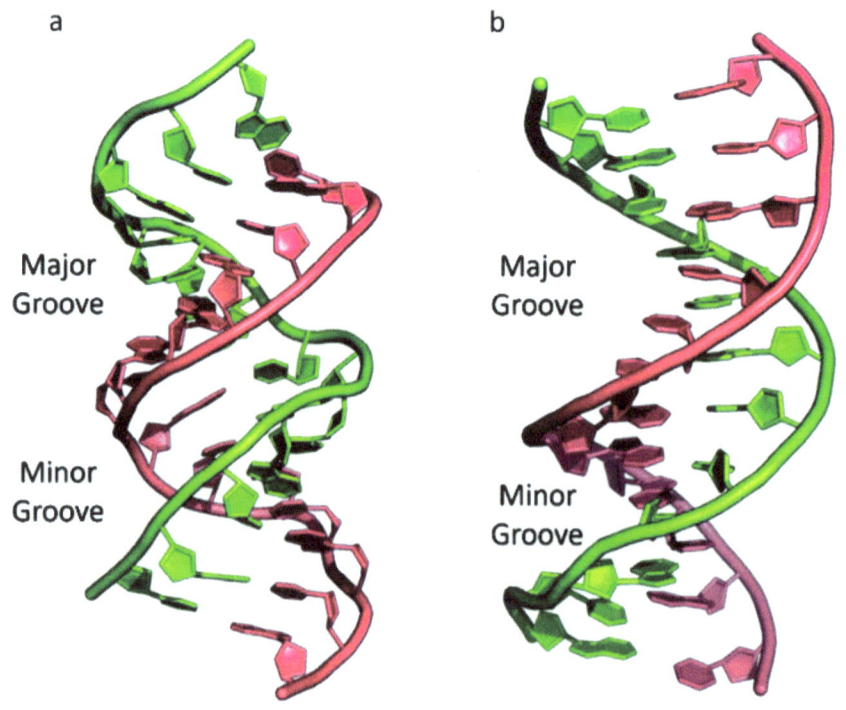

While both DNA and RNA can form triple-helix structures, novel synthetic nucleic acid analogues can also be used for TFO formation. Peptide nucleic acids (PNAs) are synthetic compounds with a neutral polyamide rather than charged phosphodiester backbone. They are more resistant to protease and nuclease degradation and can bind more tightly to DNA and RNA.

Gamma PNA molecules, which feature a pre-organized conformation, have increased binding to target DNA. Mini-PEG or other modifications may also increase PNA binding to target DNA. Locked nucleic acids (LNAs) are another synthetic oligonucleotide, with a bridge between the 2' oxygen and 4' carbon, "locking" the ribose in a fixed 3'-endo configuration. This rigid conformation reduces barriers to binding by lowering the binding entropy.

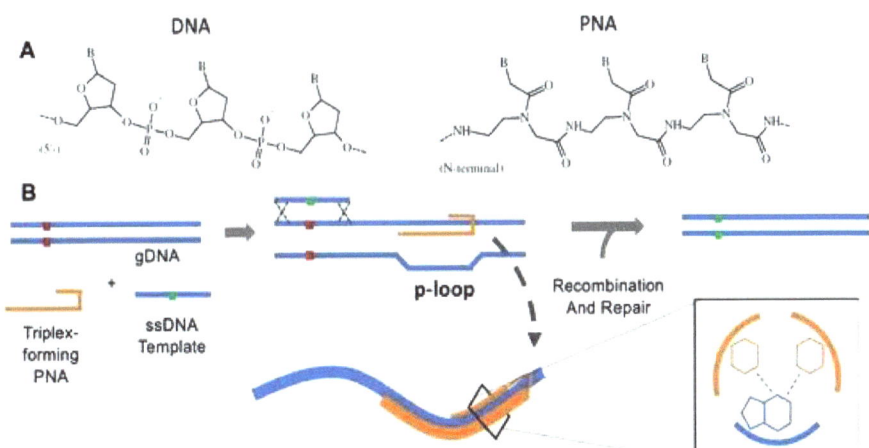

PNAs and other molecules are capable of forming unique structures with DNA. These structures include triplexes consisting of PNA binding with DNA through Hoogsteen base pairs at homopurine/pyrimidine stretches.

TFO binding has been shown to inhibit transcription, replication, and protein binding to DNA. In addition, TFOs tethered to mutagens can promote DNA damage in a sequence-specific fashion and induce mutagenesis. More recently, researchers have demonstrated that TFOs can mediate site-specific gene modification, both in vitro and in vivo.

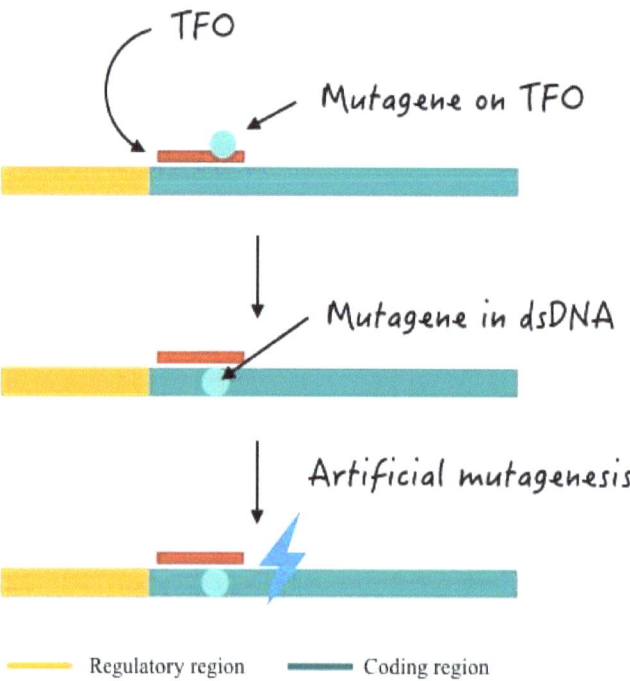

In this chapter we focus on the use of triplex-forming molecules to mediate gene modification and novel methods for TFO delivery that can be used for transfer of diverse nucleic acids. While introduction of an oligonucleotide homologous to a target gene may lead to recombination at low levels, use of TFOs can enhance recombination frequencies, leading to targeted, specific editing of endogenous human genes.

Precision Medicine: Disease Subtyping and Tailored Treatment

Wang RC, Wang Z.

Cancers (Basel). 2023 Jul 28;
15(15):3837.
doi: 10.3390/cancers15153837.
PMID: 37568653; PMCID: PMC10417651.

Excerpts

The genomics-based concept of precision medicine began to emerge following the completion of the Human Genome Project. In contrast to evidence-based medicine, precision medicine will allow doctors and scientists to tailor the treatment of different subpopulations of patients who differ in their susceptibility to specific diseases or responsiveness to specific therapies. The current precision medicine model was proposed to precisely classify patients into subgroups sharing a common biological basis of diseases for more effective tailored treatment to achieve improved outcomes.

Precision medicine has become a term that symbolizes the new age of medicine. In this review, we examine the history, development, and future perspective of precision medicine. We also discuss the concepts, principles, tools, and applications of precision medicine and related fields. In our view, for precision medicine to work, two essential objectives need to be achieved. First, diseases need to be classified into various subtypes. Second, targeted therapies must be available for each specific disease subtype. Therefore, we focused this review on the progress in meeting these two objectives.

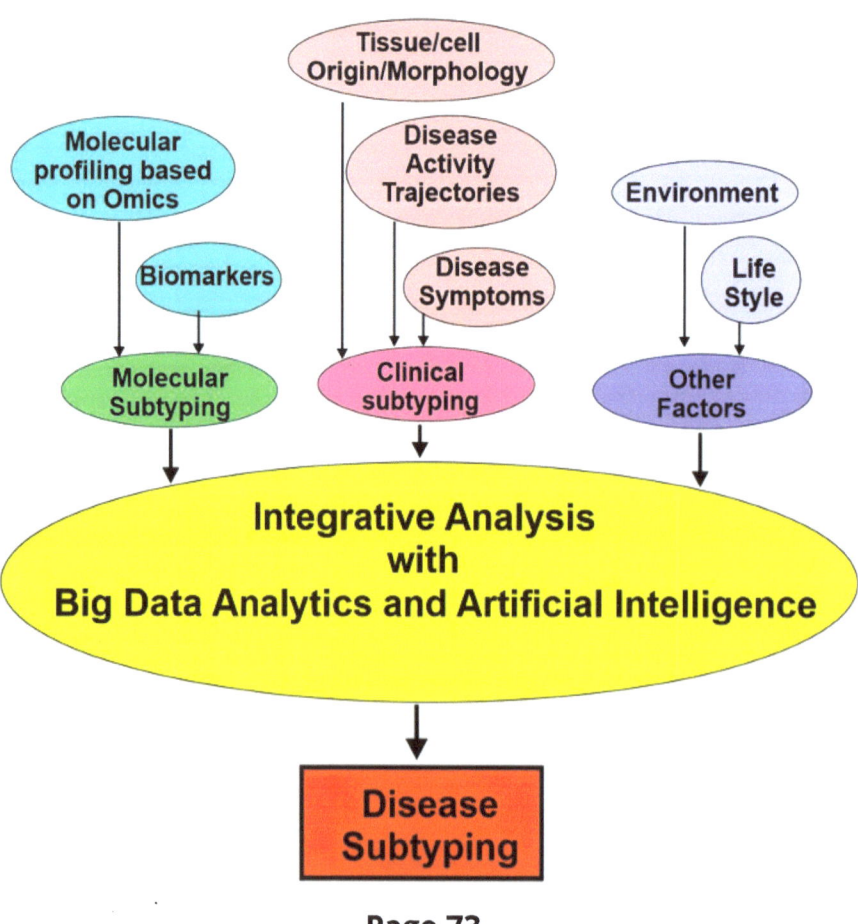

In 2011, The genomics-based concept of precision medicine began to emerge following the completion of the Human Genome Project. The US National Research Council defines precision medicine as "an emerging approach for disease treatment and prevention that takes into account individual variability in genes, environment, and lifestyle for each person".

This approach is in contrast to a one-size-fits-all approach and will allow doctors and scientists to tailor treatment to different subpopulations of people based on their unique disease susceptibility and/or treatment response.

Often, the term "precision medicine" is used synonymously with similar terms like personalized medicine, stratified medicine, individualized medicine, tailored medicine, and P4 medicine. All these terms were recently developed to contrast the traditionally used "evidence-based medicine", which in turn is the advancement from "traditional medicine".

The term precision medicine may be relatively new, but its concept has existed for a long time. Hippocrates (460-370 BCE) once famously said that "It's far more important to know what person the disease has than what disease the person has". There are numerous medical practices in existence that share the concept of precision medicine. One great example is blood typing for blood transfusions.

Due to the recent development of a large-scale biological data base and computational tools used to analyze the large sets of data, a more personalized and precise approach to medicine is becoming possible.

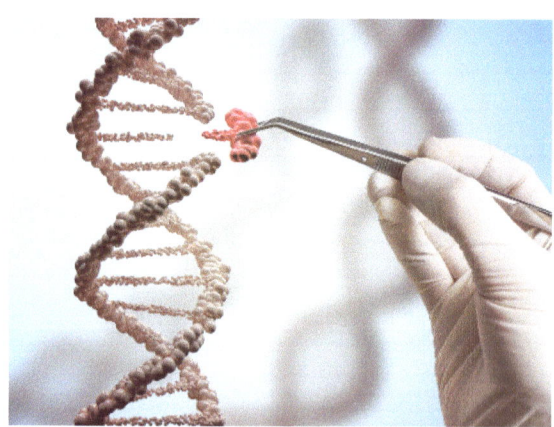

Precision medicine has often been described as the "right drugs for the right patients at the right time". To achieve this goal, the current "precision medicine" model is proposed to precisely classify patients into subgroups sharing a common biological basis of diseases for more effective treatment and improved care outcomes.

As an essential objective of precision medicine, disease subtyping is the task of classifying a disease into distinct patient subgroups based on specific patient characteristics, which can then guide treatment decisions based on which subgroup a patient belongs to.

Although disease subtyping has been conducted for many years, it was conducted as a by-product of clinical experience in past times.

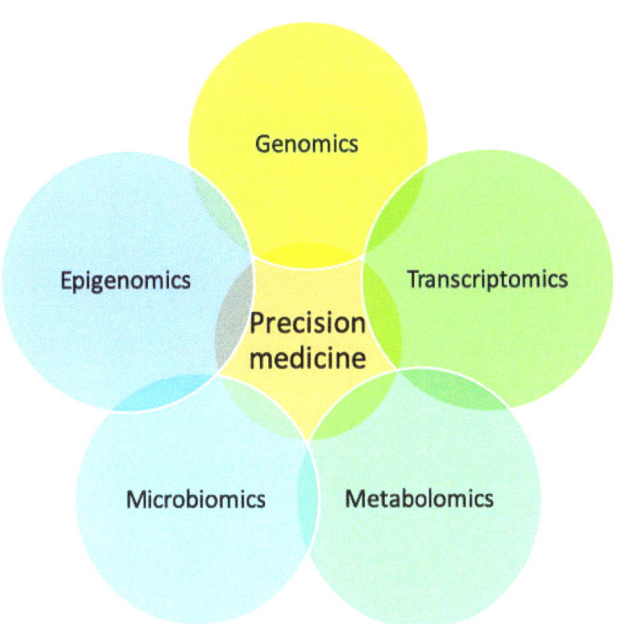

Things changed in the last two decades. With the rapid development of modern high-throughput biotechnologies, we can now measure differences between individuals at the cellular and molecular levels with various "-omics". The cost of measuring these "-omic" data (such as genomics, proteomics, and metabolomics data) has decreased significantly, which allows the collection of such data on a large number of patients.

High-throughput technologies have revolutionized medical research since the completion of the Human Genome Project at the beginning of the millennium. These new technologies can generate a high-throughput and high-resolution snapshot of any biological system of interest, which constitutes "Omics".

Omics is a nomenclature broadly used to describe a collective study of molecular characterization and the quantification of biological molecules via high-throughput technology. Based on the subdomains the molecules belong to, these omics were described as genomics, epigenomics, transcriptomics, proteomics, and metabolomics, among others.

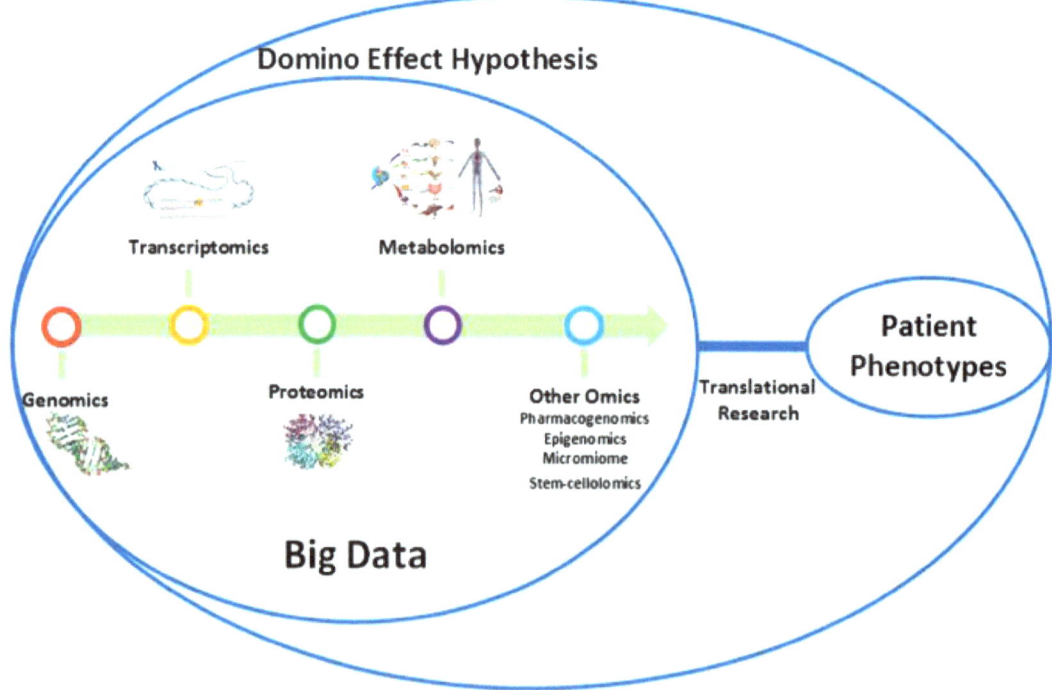

The most relied-upon high-throughput technologies in omics is next-generation sequencing (NGS), which sequences millions of individual DNA fragments simultaneously to generate a massive amount of molecular data with greater efficacy, speed, and cost-effectiveness. NGS has benefited almost all subdomains of omics research.

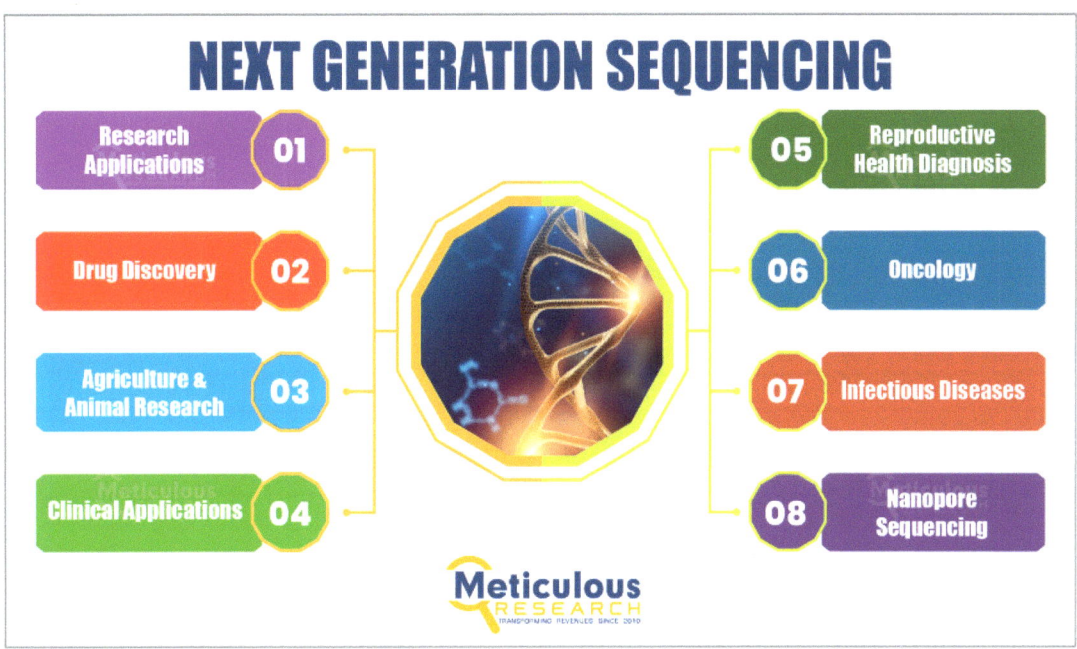

Generation and analyses of high-throughput molecular data through various omics allow the unbiased biomedical discovery of disease subtypes via the unsupervised clustering of either individual or multiple sources of molecular data.

Genomics focuses on the study of entire genomes and is the first omics discipline to appear. This is different from "genetics", which only studies individual variants or single genes. Genomic studies interrogate genetic variations, contributing to both mendelian and complex diseases. Important approaches in genomics include genome-wide association studies (GWASs), whole-genome sequencing (WGS), and whole-exome (protein-coding regions of genes) sequencing (WES).

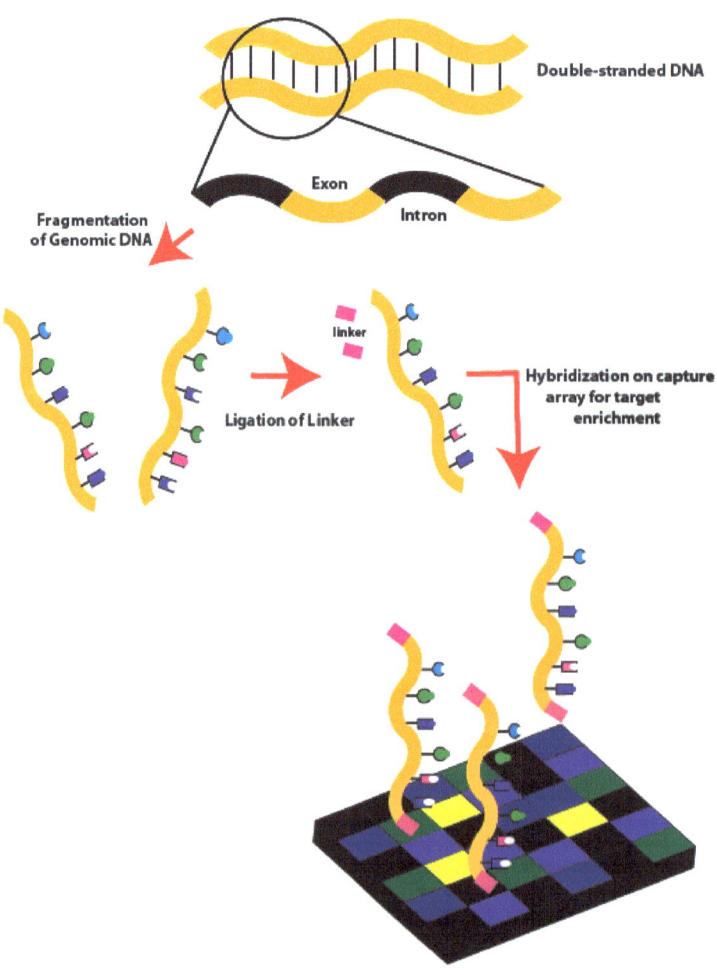

Whole-exome Sequencing (WES).

Genome-wide association studies (GWASs) have been successively used to identify thousands of genetic variants associated with complex diseases in various human populations. In GWASs, more than a million genetic markers are analyzed for a large group of individuals. Any differences in minor allele frequencies (such as single nucleotide polymorphisms or SNPs), if statistically significant between patients and controls, are treated as the evidence of association. GWASs have contributed significantly to our understanding of complex diseases.

Genome-wide assocation studies (GWAS)

Whereas GWASs aim to identify associations between genetic variants and diseases, WGS and WES focus on sequencing the genome. WGS aims to sequence the complete genome of an individual, including both coding and noncoding regions, while WES focuses specifically on sequencing the exome or the coding segments of the genome actually transcribed. Both WGS and WES rely on NGS to generate the vast amounts of data needed.

Another powerful genomics tool to reveal cellular complexity is single-cell genomic sequencing (SGS), which focuses on analyzing the genome of individual cells. This is highly useful for exploring individual genomic variation (mosaicism), such as when differentiating between tumor and nontumor cells.

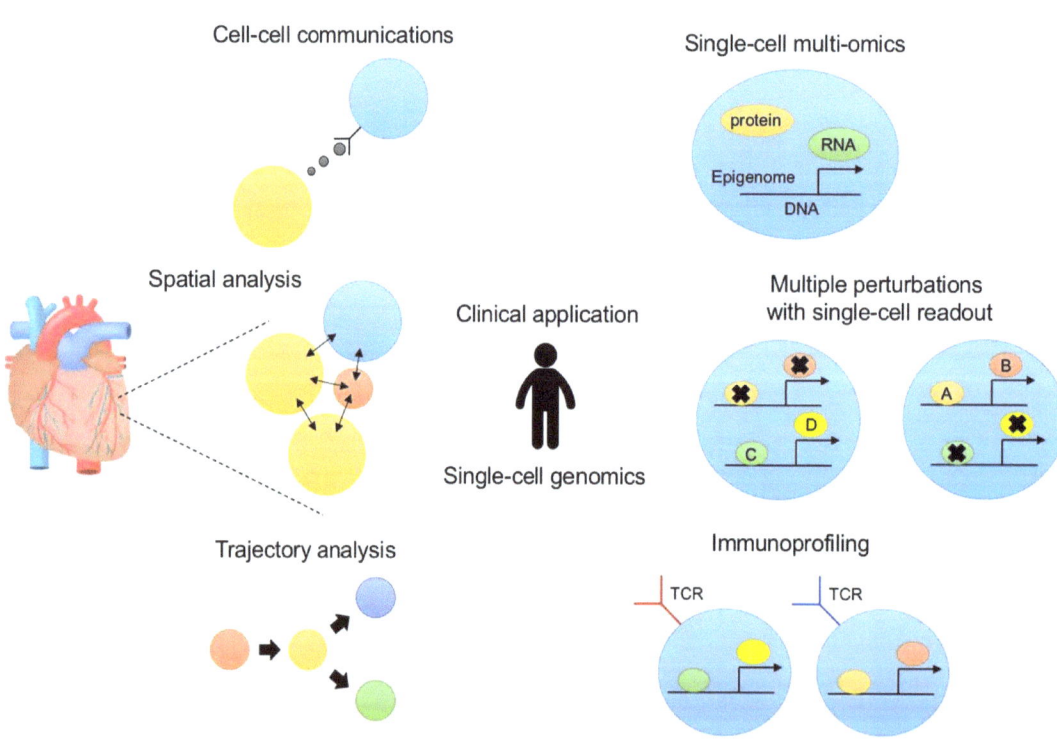

Pharmacogenetics and pharmacogenomics investigate the effects of genetic variations on drug responses and dispositions. Genetic variation may occur in single genes or in multiple genes (polygenetic). If the frequency of a genetic variation is more than 1% of the population, it is defined as a genetic polymorphism. Originally, pharmacogenetic studies focused on monogenic variations that are involved in drug transportation and metabolism.

Pharmacogenetics
- Reactive observational studies
- Family-based studies
- Candidate gene studies
- Randomized controlled trials

Pharmacogenomics
- Genome-wide association studies
- High-throughput sequencing
- Population comparative studies
- High-throughput functional validation

Pharmaco-omics
- Electronic health records
- Transcriptome, proteome, metabolome, epigenome
- Machine learning
- *In silico* approaches

The goal of pharmacogenetics is to make an informed and best choice of medical treatment based on a patient's individual genetic information. However, with advances in molecular pharmacology and human genomics in the late twentieth century, pharmacogenetics evolved into the more comprehensive pharmacogenomics.

Both pharmacogenetics and pharmacogenomics support precision medicine by identifying drugs tailored to the individual's particular genetic makeup. Although a person's environment, state of health, lifestyle, age, and diet can also influence that person's response to medicines, an individual's genetic makeup is the key to selecting personalized drugs that work best for the patients with fewer side effects.

To realize the full promise of precision medicine, a further integration of multiple levels of information is needed. This includes the integration of information at the molecular level (i.e., genome, epigenome, proteome, metabolome), clinical and laboratory data, and environmental factors. Big data analytics and artificial intelligence will be the key to accomplishing this integration.

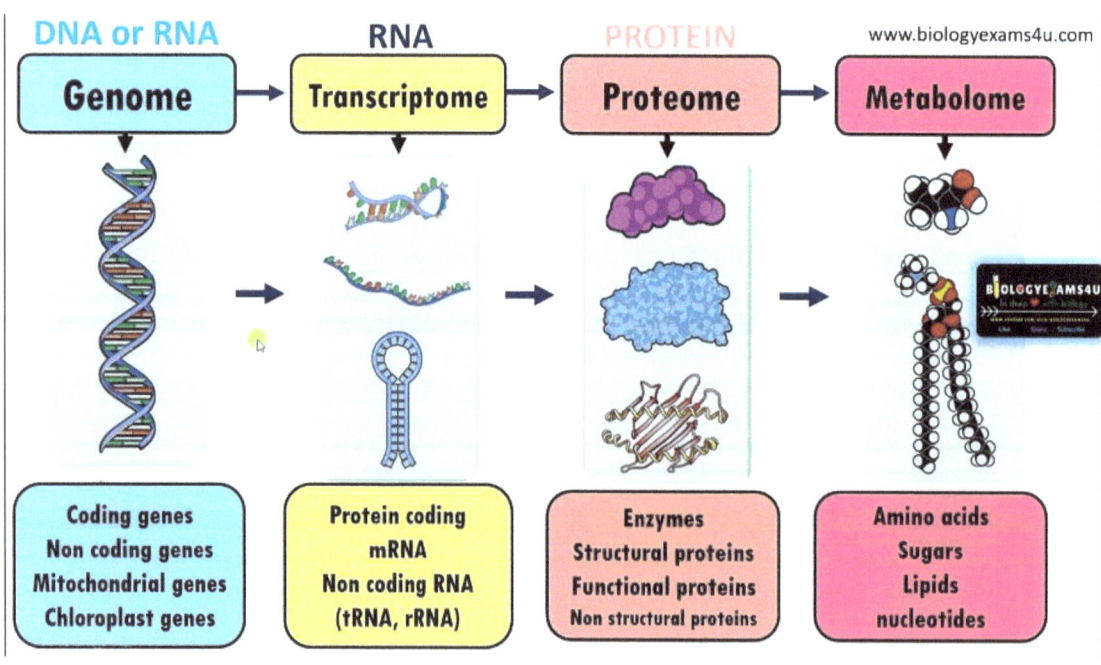

PLEASE JOIN US!

Kathleen and I conduct three livestreams each week on Tuesday, Thursday, and Sunday from 7:00PM - 9:00PM Eastern. In addition, we facilitate a communication platform through BAND.

We are completely private and no educational livestream videos are posted to any other mainstream platform. We are a subscription-based research provider and educational speakers.

Please consider joining us. Some membership levels include a subscription to Entangled magazine.

https://www.subscribestar.com/anthony-patch

For more information, please visit: https://www.anthonypatch.com